MECHANICS OF RESIDUAL SOILS

Mechanics of Residual Soils

A guide to the formation, classification and geotechnical properties of residual soils, with advice for geotechnical design

Edited by
G.E. BLIGHT
Chairman of Technical Committee 25, International Society for Soil Mechanics and Foundation Engineering
Department of Civil Engineering, Witwatersrand University, Johannesburg, South Africa

Prepared by Technical Committee 25 on The Properties of Tropical and Residual Soils of the International Society for Soil Mechanics and Foundation Engineering

BALKEMA/ROTTERDAM/BROOKFIELD/1997

Published by
A.A.Balkema, P.O.Box 1675, 3000 BR Rotterdam, Netherlands
Fax:+31.10.4135947; E-mail: balkema@balkema.nl; Internetsite: http://www.balkema.nl

A.A.Balkema Publishers, Old Post Road, Brookfield, VT 05036-9704, USA
Fax: 802.276.3837; E-mail: info@ashgate.com

ISBN 90 5410 696 4

Contents

CHAPTER 1

Origin and formation of residual soils

G.E. BLIGHT
Department of Civil Engineering, Witwatersrand University, Johannesburg, South Africa

The definition of a residual soil varies from country to country, but a reasonably general definition would be:

'A residual soil is a soil-like material derived from the in situ weathering and decomposition of rock which has not been transported from its original location'.

There may be a continuous gradation from the fresh, sound unweathered rock through weathered soft rock and hard soil, which is recognizable as the product of decomposition of the parent rock, to highly weathered material containing secondary deposits of iron, alumina, silica or calcium salts that bears no obvious resemblance to the parent material.

Residual soils can have characteristics that are quite distinctively different from those of transported soils. For example, the conventional concept of a soil grain or a particle size is inapplicable to many residual soils. Particles of residual soil often consist of aggregates or crystals of weathered mineral matter that break down and become progressively finer if the soil is manipulated. What appears in situ to be a coarse sandy gravel may deteriorate to a fine sandy silt during excavation, mixing and compaction. The permeability of a transported soil can usually be related to its granulometry (e.g. by the well-known Hazen formula). For the reason just explained, this may not be so with residual soils. The permeability of residual soils is usually governed by its micro- and macro-fabric and jointing and by superimposed features such as slickensiding, termite or other bio-channels.

1.1 WEATHERING PROCESSES

Residual soils are formed by the in situ weathering of rocks, the three major agencies of weathering being physical, chemical and biological processes. In the weathering process the parent rock and rock minerals break down, releasing internal energy and forming substances having a lower internal energy which are therefore more stable. Physical processes (e.g. stress release by erosion, differential thermal strain and ice and salt crystallization pressures) comminute the rock, expose fresh surfaces to chemical attack and increase the permeability of the material to the percolation of

1

chemically reactive fluids. Chemical processes, chiefly hydrolysis, cation exchange and oxidation alter the original rock minerals to form more stable clay minerals (Mitchell 1976). Biological weathering includes both physical action (e.g. splitting by root wedging) and chemical action (e.g. bacteriological oxidation, chelation and reduction of iron and sulphur compounds (e.g. Pings 1968).

Most commonly, residual soils form from igneous or metamorphic parent rocks, but residual soils formed from sedimentary rocks are not uncommon. Chemical processes tend to predominate in the weathering of igneous rocks, whereas physical processes dominate the weathering of sedimentary and metamorphic rocks. However, chemical and physical weathering are so closely interrelated that one process never proceeds without some contribution by the other.

Occasionally, residual soils may form by the in situ weathering of unconsolidated sediments. The commonest example of this is the loess or collapsing sand formed by the weathering of feldspars in deposits of windblown sand (Knight 1961, Dudley 1970, Schwartz & Yates 1980). Shirasu, found in Japan (Yamanouchi & Haruyama 1969) is an unconsolidated volcanic sediment, partly weathered in situ, whose engineering properties have much in common with loess. A survey by Brand & Phillipson (1985) showed that colluvium mantling a residual soil profile is regarded as a residual soil by many authorities, even though it has been transported a short distance from its place of origin.

Hydrolysis is considered to be the most important of the chemical weathering processes (Zaruba & Mencl 1976). It occurs when a salt combines with water to form an acid and a base. In rock weathering, the salt is usually a silicate and the product of the reaction is a clay mineral. Oxidation is usually preceded by hydrolysis and affects rocks containing iron sulphates, carbonates and silicates. The products of oxidation usually have a larger specific volume than the parent minerals and thus expansion due to oxidation contributes to physical comminution of rock (e.g. Mason 1949).

The breakdown of one clay mineral to form another can occur through the transfer of ions between percolating solutions and the original mineral. Cations such as sodium and calcium are the most readily exchangeable. Cation exchange does not alter the basic structure of the clay mineral, but the crystal interlayer spacing may change thus converting, for example, an illite to a montmorillonite.

Bacteria may play the role of catalysts in certain chemical reactions. For example, the oxidation of sulphide minerals may be enormously accelerated by the presence of the bacteria thio-bacillus thiooxidans and thiobacillus ferro-oxidans.

Chelation is a process whereby lichens growing on rock surfaces promote the rate of hydrolysis. Jackson & Keller (1970) have shown that the depth of weathering of basalt surfaces is greater and the chemical alteration more extensive under a lichen cover than if lichen is not present.

1.2   CLIMATE

Climate exerts a considerable influence on the rate of weathering (Weinert 1964, 1974, Morin & Ayetey 1971). Physical weathering is more predominant in dry climates while the extent and rate of chemical weathering is largely controlled by the availability of moisture and by temperature. (Other things being equal, chemical re-action rates approximately double for each 10°C rise in average temperature).

According to Uehara (1982) the clay mineralogy of the soils of the world changes in a predictable way with distance from the equator, as indicated by Figure 1.1. This is a gross over-simplification because as shown by Figure 1.2, climates do not vary uniformly with distance from the equator, but are affected by topography, ocean currents, etc. Nevertheless, Figure 1.1 gives a useful concept of the influence of climate on the products of weathering. Uehara says that near the equator, high temperatures and year-round rainfall favour the formation of low activity kaolin and oxides (Skempton's Activity 0.3 to 0.5). As one moves towards the limits of the tropics, rainfall decreases and high activity smectitic clays predominate (Skempton's Activity 1.5 to 7). The above concept has been presented far more elaborately by Strakhov (1967), whose diagram (Fig. 1.3) summarizes the effects of global climate on rock weathering and the formation of the various weathering products – kaolinite, alumina, etc. The influence of temperature and moisture on rock weathering in South Africa has been correlated with Weinert's (1974) climatic index

$$N = \frac{12Ej}{Pa} \tag{1.1}$$

where Ej = evaporation during January, the hottest southern hemisphere month, and Pa = the annual rainfall.

A value of $N = 5$ marks the transition from warm sub-humid conditions in which chemical weathering predominates to hot semi-arid and arid conditions in which physical weathering is the more important mode. Where N is less than 5, considerable thicknesses of residual soil may occur. Where N exceeds 5, thicknesses of residual material are usually small. Figure 1.4 shows the sub-division of Southern Africa by the Weinert's $N = 5$ contour. In the area to the east and south, chemical weathering predominates, while to the west and north, physical weathering is predominant.

Climate has a further possible effect on the properties of tropical residual soils – that of unsaturation. Even in sub-humid tropical or subtropical areas, water tables are often deeper than 5 to 10 m and the effects of unsaturation, desiccation and seasonal or longer term re-wetting have to be taken into account in geotechnical design. There

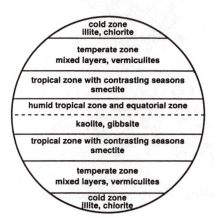

Figure 1.1. Suggested influence of global climate on clay mineral development (Uehara, 1982).

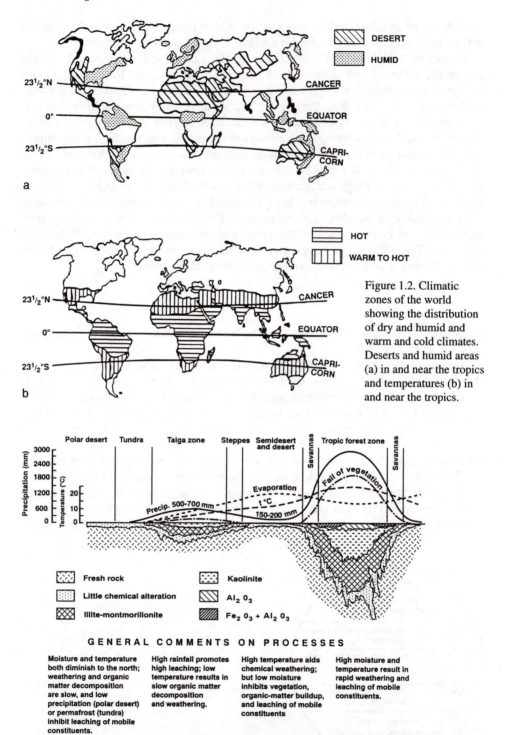

Figure 1.2. Climatic zones of the world showing the distribution of dry and humid and warm and cold climates. Deserts and humid areas (a) in and near the tropics and temperatures (b) in and near the tropics.

GENERAL COMMENTS ON PROCESSES

Moisture and temperature both diminish to the north; weathering and organic matter decomposition are slow, and low precipitation (polar desert) or permafrost (tundra) inhibit leaching of mobile constituents.

High rainfall promotes high leaching; low temperature results in slow organic matter decomposition and weathering.

High temperature aids chemical weathering; but low moisture inhibits vegetation, organic-matter buildup, and leaching of mobile constituents

High moisture and temperature result in rapid weathering and leaching of mobile constituents.

Figure 1.3. Influence of global climate on depth of weathering and weathering products (Strakhov 1967).

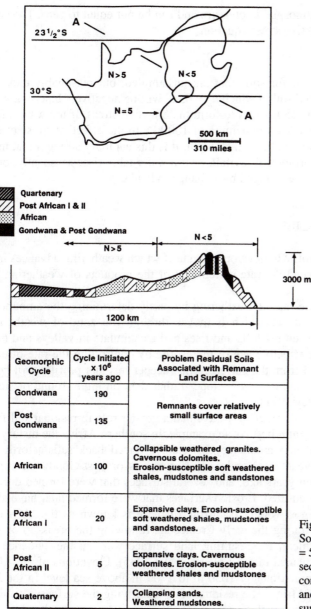

Figure 1.4. Subdivision of Southern Africa by Weinert's N = 5 contour (a), and transverse section AA showing sub-continental topographical relief and location of ancient erosion surfaces (b).

Geomorphic Cycle	Cycle Initiated x 10^6 years ago	Problem Residual Soils Associated with Remnant Land Surfaces
Gondwana	190	Remnants cover relatively small surface areas
Post Gondwana	135	
African	100	Collapsible weathered granites. Cavernous dolomites. Erosion-susceptible soft weathered shales, mudstones and sandstones
Post African I	20	Expansive clays. Erosion-susceptible soft weathered shales, mudstones and sandstones.
Post African II	5	Expansive clays. Cavernous dolomites. Erosion-susceptible weathered shales and mudstones
Quaternary	2	Collapsing sands. Weathered mudstones.

are many accounts of the effects of unsaturation on the behaviour of soils (e.g. Bishop & Blight 1963, Blight 1967, Fredlund & Morgenstern 1977, Blight 1983, Fredlund & Rahardjo 1993) and the best form of the effective stress relationship for unsaturated soils is not yet agreed. However, it is agreed that the effective stress is governed by the stress difference $(\sigma - u_a)$ and the suction $(u_a - u_w)$ where u_a and u_w are, respectively the pore air and pore water pressures in the soil. In most practical situations, u_a equals the atmospheric pressure and can be put equal to zero. The conventional form of the effective stress equation

$$\sigma' = \sigma - u_w \tag{1.2}$$

can be used with little error for soils that are unsaturated, but reasonably close to saturation. In an unsaturated soil in which $u_a = 0$, u_w will be negative. Hence the pore water stress is added to the total stress to form the effective stress. If the water table is at a depth of 10 m, this adds a maximum of 100 kPa to the effective stress in the soil profile and about 50 kPa to the soil strength. It is this capillary strength that may be lost during periods of prolonged rainfall, or if a water table rises seasonally, or if conditions of unsaturation are changed by a change in land use.

1.3 TOPOGRAPHIC RELIEF

For a deep residual soil profile to develop, the rate at which weathering advances into the earth's crust must exceed the rate of removal of the products of weathering by erosion.

Topography controls the rate of weathering by partly determining the amount of available water and the rate at which it moves through the zone of weathering (Precipitation will tend to run off hills and rises and accumulate in valleys and hollows). It also controls the effective age of the profile by controlling the rate of erosion of weathered material from the surface. Thus deeper residual profiles will generally be found in valleys and on gentle slopes rather than on high ground or steep slopes (e.g. Morin & Ayetey 1971).

It thus follows that ancient relief may have had a greater role in residual soil formation than does more recent relief. As an example, in southern Africa the history of land surface development and erosion cycles can be traced back satisfactorily to Gondwana (190 million years ago) and early Cretaceous or post-Gondwana times (approximately 135 million years ago, and several surfaces that were formed during these periods have been identified. Erosion surfaces that were formed later have also been identified, and the most widely occurring of these is known as the 'African' erosion surface. Initiated during the early Cretaceous following the break-up of the Gondwanaland supercontinent, the resulting land surface was, on the evidence of surviving remnants, well planed in most places (Falla 1985). Elevations of the African surface range from approximately 1650 m to 1700 m above sea level in the Johannesburg area of South Africa. The residual soils underlying the surface have distinctive engineering characteristics owing to the great age of the surface (approximately 100 million years).

In the Johannesburg area there are several remnants of the African surface, but erosion has formed several Post-African variants of the original surface. One variant that has suffered widespread lowering and stripping of residual cover is described as the 'Lowered African' surface. Conversely, surfaces that lie above the general level of the African surface also occur. These have suffered a similar degree of weathering since 'African' times. This zone, which is not a planed surface, is characterized by undulating topography, and is described as the 'Above African' surface. These three distinct surfaces or zones are close in geological age, but decrease in age from the

Above African to the African to the Lowered African. As will be shown later, there are distinct differences in the characteristics of the soil profile that underlies each surface. Figure 1.4 shows the elevation and location of the ancient erosion surfaces that have been identified in Southern Africa, together with a tabulation of the age of the surfaces and the geotechnical problems associated with soils beneath each surface.

Fitzpatrick & Le Roux (1977) studied soil profiles developed from basic igneous rock on hillsides. They found that the depth of weathering increased down the slope. Whereas kaolinite/halloysite were the predominant clay minerals at the top of the slope (see Fig. 1.5) that at the bottom of the slope was smectite.

Van der Merwe (1965) made a study of residual soils derived from a number of basic igneous rock types. In a study of weathered diabase profiles from three different sites he found that the predominant factors, other than climate and rainfall, affecting development of clay minerals were local topography and internal drainage. Samples taken from a site high upon a slope with good runoff showed kaolinite and vermiculite to be the dominant clay minerals. A flatter site indicated a chlorite, vermiculite, montmorillonite and kaolinite weathering sequence. The third site was flat, with impeded drainage. Because of leaching of alkalies and alkali earths and predominantly reducing conditions, weathering could not proceed to the kaolinite stage and hence montmorillonite was the predominant mineral found. This study indicated that good internal drainage and high rainfall are favourable to the development of kaolinite whilst flatter slopes and poor drainage favour the formation of montmorillonite.

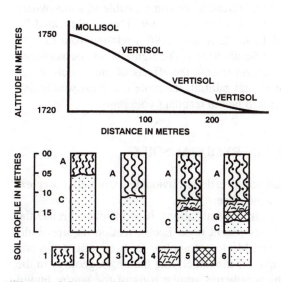

Figure 1.5. Section of the Transvaal black clay toposequence (the cover consists of grassy meadow vegetation *Themeda triandra*). 1) Black clay with a strong subangular blocky structure (breaking down to a good crumb structure), 2) Black clay with a strong coarse subangular blocky structure (breaking down to a strong medium angular blocky structure) with vertical cracks and slickensides, 3) As for 2 but with a small hard round calcareous nodules, 4) Narrow moist band (approximately 6 cm) of grey friable soft calcium carbonate in a yellowish grey clay matrix, 5) Pleyed clay with slickensides, 6) Yellowish, olive brown friable decomposed dolerite.

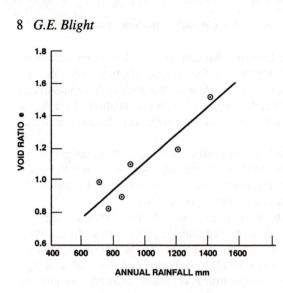

Figure 1.6. Relationship between void ratio and annual rainfall for highly weathered and leached granites in South Africa.

Topography also controls the drainage of an area. Local variations in topography influence the amount of moisture retained, the position and configuration of the water table, and thereby the depth of penetration of most chemical weathering processes. The relief of the area thus controls whether a soil develops under conditions of poor or good drainage.

Under conditions of good drainage, topography controls the extent of leaching, an essential mechanism of rock weathering. Highly weathered residual granites, for instance are found to be more porous and therefore more compressible in zones where the annual precipitation and therefore the leaching is greater. This is illustrated by Figure 1.6 which shows some data of Brink & Kantey (1961) relating void ratio to annual rainfall for weathered granites in South Africa. The figure is, of course, based on current rainfall patterns and therefore represents the effects of the climate in recent times. Obviously, the higher the annual rainfall, the more chemical weathering and leaching will occur, and the higher will be the resultant void ratio.

1.4 GENERAL CHARACTERISTICS OF RESIDUAL SOILS

The process of formation of a residual soil profile is obviously extremely complex, difficult to understand and difficult to generalise.

It is evident that apart from a few valid generalisations, it is difficult to relate the properties of a residual soil directly to its parent rock. Each situation requires individual consideration and it is rarely possible to extrapolate from experience in one area to predict conditions in another, even if the underlying hard rock geology in the two areas is similar. For instance, the weathered granite soils of the warm, humid Malaysian peninsular have very different properties to those of the cooler, semi-arid Transvaal Highveld in South Africa.

The chemical changes and sequences of minerals formed during weathering are extremely complex. For example, one suggested weathering sequence leading to the formation of clay minerals is shown in Figure 1.7a (van der Merwe 1965). The sequence may be arrested at any stage and it is possible that certain stages may be reversed as a result of changes in climate or conditions of drainage.

Figure 1.7b (Gonzalez de Vallejo et al. 1981) shows weathering sequences for volcanic rock observed in Cameroun (West Africa) and Kenya (East Africa). Apart from unknown differences in mineralogy, between the parent rock materials, Cameroun has a hotter, moister climate than does Kenya, which may partly explain the difference in weathering sequence.

As mentioned previously, and indicated in Figure 1.7a, van der Merwe (1965) has shown that reddish kaolinitic soils develop in well drained situations over norite gabbro, whereas blackish montmorillonitic clays develop from identical parent rock in poorly drained situations, on the South African Highveld.

Wesley (1973) concluded that the dark-coloured andosols and red latosols found in Java originate from much the same volcanic parent material but occur in profiles of different ages.

Because weathering proceeds from the surface down and inwards from joint surfaces and other percolation paths, the intensity of weathering generally reduces with increasing depth and reducing intensity of jointing in the material between joint surfaces. In profiles residual from igneous rocks, core stones or boulders of sound parent rock are very often found enclosed within blocks of weathered rock (e.g. Lumb 1961). This is the typical 'onion skin' weathering pattern so often seen in basic igneous rocks such as basalts and dolerites. Typically, a profile of residual soil will con-

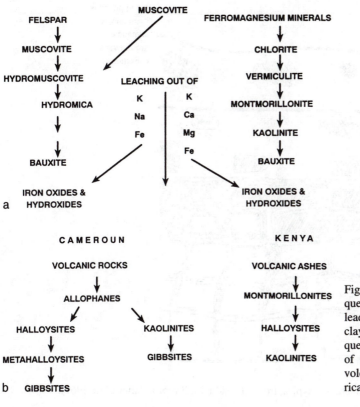

Figure 1.7. Suggested sequence of weathering leading to formation of clay minerals (a), and sequences in the formation of clay minerals from volcanic materials in Africa (b).

sist of three indistinctly divided zones (Vargas & Pichler 1957, Ruxton & Berry 1957, Little 1969) as illustrated in Figure 1.8. The upper zone will consist of highly weathered and leached soil often reworked by burrowing animals and insects or by cultivation, and intersected by root channels. This zone has always been subjected to at least some transport processes. The intermediate zone also consists of highly weathered material but exhibits some features of the structure of the parent rock and may contain core stones. This zone often contains pedogenic material such as nodules of calcium or iron salts which may give it a mottled or spotted appearance.

Saprolites are materials that have soil-like strength or consistency, but retain modified but recognisable relics of the physical features or fabric of the parent rock. For example a saprolite derived from the weathering of lava may retain the flow structure and amygdales of the parent rock. One derived from a shale often retains the bedding and jointing pattern of its parent rock. Saprolites may occur in profiles of almost any depth.

Laterites are usually highly weathered and altered residual soils, low in silica, that contain a sufficient concentration of the sesquioxides of iron and aluminium to have

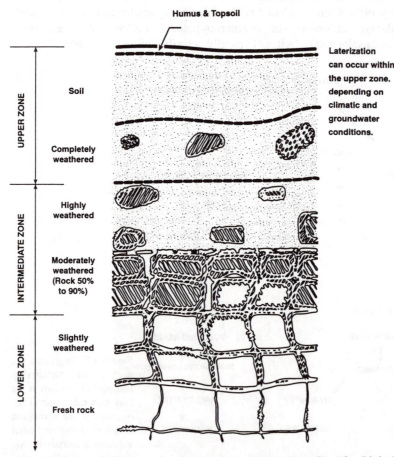

Figure 1.8. Schematic diagram of typical tropical residual soil profile (after Little 1969).

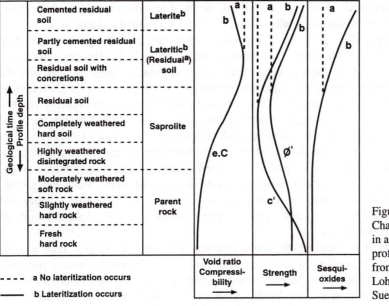

Figure 1.9. Changes occuring in a weathering profile (adapted from Tuncer & Lohnes 1977 and Sueoka 1988).

been cemented to some degree. These salts are secondary emplacements resulting from the evaporation of iron- or aluminium-bearing near-surface ground waters. Depending on the extent of the emplacement, the material could be described as lateritic or as laterite. Lateritization usually occurs in residual soils, but ancient transported soils may also have been lateritized. Desai (1985) gives definitions of the degree of lateritization in terms of the silica/alumina ratio. Unlaterized soils have SiO_2/Al_2O_3 greater than 2. For lateritic soils SiO_2/Al_2O_3 lies between 1.3 and 2 and for true laterites the ratio is less than 1.3.

Figure 1.9 (adapted from Tuncer & Lohnes (1977) and Sueoka (1988) shows in schematic form the progression from fresh rock through saprolite to laterite. Note that the progression from residual soil to laterite is not inevitable, but depends on conditions being favourable for lateritization to occur. Most of the depth of a residual profile will usually not be subject to lateritization. Because of their mode of formation, laterites usually occur as fairly near-surface strata of limited depth. Laterites are often excellent materials for the building of roads and embankments (e.g. Medina 1989, Hight et al. 1988, Sweere et al. 1988, Gidigasu 1988) and are often sought after for this purpose. They may also form good shallow founding strata of high strength and low compressibility.

1.5 AN EXAMPLE OF THE FORMATION OF A RESIDUAL SOIL PROFILE BY THE WEATHERING OF ANDESITE LAVA

This example is given to show in detail how a typical residual soil profile has developed and to describe its properties. In this chapter, the occurrence, appearance, weathering process and mineralogy of the soil will be described. In later chapters the mechanical properties of the same soil will be used to give examples of the compressibility, shear strength and other characteristics of a typical residual soil.

Andesite is part of a family of dark coloured rocks in which the predominant dark minerals are hypersthene, augite and horneblende (Ernst 1969). The composition of andesite in terms of the ratio of alkali (Sodium-Potassium) to total feldspar and the anorthite (Calcium Aluminium Silicate) content of the plagioclase (Na-Ca feldspar) as well as its relationship to other dark coloured rocks is shown in Figure 1.10. The andesite considered in this paper is part of the approximately 2500 million year old Ventersdorp Supergroup of rocks in South Africa (Brink 1979, Falla 1985).

The igneous rocks of the Ventersdorp Supergroup consist mainly of andesitic lavas with interbedded zones of reworked volcanic conglomerates, agglomerates and tuffs. The principal minerals of these rocks are feldspar and augite, occurring as fine laths with occasional phenocrysts, together with amygdales of quartz, epidote and some chlorite, while the volcanic conglomerates consist of clastic accumulations of compacted lava fragments.

The constituent minerals in the andesite have weathered to produce a usually well-defined colour sequence in the residual soil. The ferro-magnesian minerals (pyroxenes) have altered to chlorite, which imparts a distinct greenish colour to the lower zones of the residual soil. As the chlorite has altered higher in the profile, iron in the lattice structure has oxidised and hydrated to form limonite or yellow ochre, which gives the soil a yellow or yellowish brown colour. In zones above the water table, where the soil may seasonally be desiccated, haematite or red ochre has been produced, giving a characteristic red or reddish brown colour, particularly near the surface. These three colours do not always all appear in the profile, as this depends on local conditions, and the boundary between each colour zone is seldom sharply defined. The tendency for this colour sequence to develop provides a convenient means for recognition and reference to engineering and weathering characteristics.

Iron compounds often become concentrated in the upper zones of a residual soil profile to form ferricrete. This is produced by the seasonal precipitation and gradual accumulation of insoluble iron oxides and hydroxides within the soil mass. Its devel-

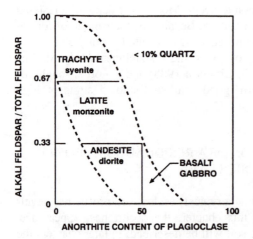

Figure 1.10. Minerological composition of andesites and relationship to other dark-coloured rocks (CAPITALS denote extrusive rocks and lower case intrusive rocks) (Ernst 1969).

opment depends on the amount of iron available and on repeated annual fluctuations in water table levels. The ferricrete may vary from thick, impermeable, compact, brownish accumulations of nodular 'hardpan' to a dark ferruginous staining in the soil. Figure 1.11 shows a typical soil profile residual from andesite in the Johannesburg area of South Africa.

As stated earlier, the formation of a profile of residual soil depends among other factors, on the rate of erosion of the surface. Obviously, if the rate of removal of material from the surface equals or exceeds the rate of advance of the 'weathering front' into the rock, no residual soil will form, and vice versa. It would therefore be expected that in a relatively small area over which climatic conditions are uniform, more deeply weathered profiles will occur beneath older surfaces. Falla (1985) showed that this is indeed so for the andesite profile being examined here. His statistical data shows some interesting trends, that can be summarised as follows:

– The total thickness of residual soil decreases progressively from that beneath the more than 100 million year old 'Above African' surface (where it averages 59 m) to that beneath the less than 100 million year old 'Lowered African' surface (where it averages only 16 m).

– The predominant soil colour in the older profiles is red, while in the younger profiles it is yellow.

The statistics of the geotechnical parameters for these soils do not show very clear trends, but this is because the whole range of degrees of weathering was present in most of the test holes used for compiling the data. However, the data do show that residual andesite soils can vary from clayey soils (clay content approaching 50% and Plasticity Index above 50) to silty soils (clay content less than 10% and Plasticity Index less than 10). Void ratios can be high (greater than 1.5) to moderate (0.6), and

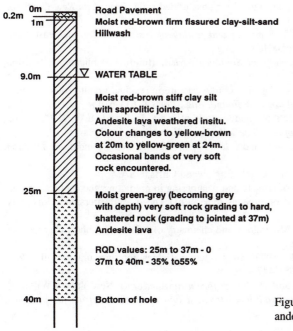

Figure 1.11. Typical profile of residual andesite lava.

compressibilities likewise can vary from C_c of over 0.9 to less than 0.1. More details of geotechnical properties will be given in later chapters.

1.6   RAPIDITY OF WEATHERING

Although it is usually assumed that soil-forming processes such as the decomposition of igneous rocks and the formation of ferricrete take place very gradually over time spans of millions of years, there is evidence that this is not always the case.

Lava flows on Hawaii island, for example, show signs of physical and chemical weathering within a few years of deposition and soon support pockets of grass and small shrubs. Ammunition buried in the soil during the South African War of 1899-1902 has been found completely encased in lateritic material 90 years later. On the other hand, ancient stone monuments such as Stonehenge in England and the Nuragi of Sardinia show very little sign of weathering after 5000 years of exposure to the elements.

The very rapid disintegration or slaking of certain mudrocks and lavas on exposure to the atmosphere is another example of weathering occurring over a time period of a few years or even months.

REFERENCES

Bishop, A.W. & Blight, G.E. 1963. Some aspects of effective stress in saturated and partly saturated soils. *Géotechnique* 13(3): 177-197.

Blight, G.E. 1967. Effective stress evaluation for unsaturated soils. *Jour, Soil Mech. & Found. Eng. Div., ASCE* 93(SM2): 125-148.

Blight, G.E. 1983. Aspects of the capillary model for unsaturated soils. *Proc. Asia Reg. Conf. Soil Mech. & Found. Eng.*, Haifa, Israel.

Brand, E.W. & Phillipson, H.B. 1985. *Sampling and testing of residual soils*. Southeast Asian Geotechnical Society, Scorpion Press, Hong Kong.

Brink, A.B.A. 1979. *Engineering geology of Southern Africa*. Building Publications, Pretoria, South Africa.

Brink, A.B.A. & Kantey, B.A. 1961. Collapsible grain structure in residual granite soils in Southern Africa. *Proc. 5th Int. Conf. Soil Mech. & Found. Eng., Paris* 1: 611-614.

Desai, M.D. 1985. Geotechnical aspects of residual soils of India. In Brand, E.W. & Phillipson, H.B. (eds), *Sampling and Testing of Residual Soils*. Hong Kong: Scorpion Press.

Dudley, J.H. 1970. Review of collapsing sands. *Jour. of Mechanics and Foundations Division, ASCE* 96(SM3): 925-947.

Ernst, W.G. 1969. *Earth Materials*. Prentice Hall, Englewood Cliffs, USA.

Falla, W.J. 1985. On the significance of climate and weathering in predicting geotechnical characteristics of residual soils developed on igneous rocks. PhD Thesis, Witwatersrand University, Johannesburg, South Africa.

Fitzpatrick, R.Q. & le Roux, J. 1977. Mineralogy and chemistry of a Transvaal black clay topo sequence. *Jour. Soil Sci.* 28: 165-179.

Fredlund, D.G. & Morgenstern, N.R. 1977. Stress state variables for unsaturated soils. *Jour. Geotech. Eng. Div., ASCE* 103(GT5): 447-466.

Fredlund, D.G. & Rahardjo, H. 1993. *Soil Mechanics for unsaturated soils*. New York: J.Wiley.

Gidigasu, M.D. 1988. *The use of non-traditional tropical and residual materials for pavement construction,* pp. 397-404.

Gonzalez de Vallejo, L.I., Jiminez Salas, J.A. & Leguey Jiminez 1983. Engineering geology of the tropic volcanic soils of La Laguna, Tenerif. *Eng. Geol.* 17(1): 1-17.

Hight, D.W., Toll, D.G. & Grace, H. 1988. *Naturally occurring gravels for road construction,* pp. 405-412.

Jackson, T.A. & Keller, W.D. 1970. A comparative study of the role of lichens and inorganic processes in the chemical weathering of recent Hawaiian laval flows. *Amer. Jour. Sci.* 269: 446-466.

Knight, K. 1961. The collapse of structure of sandy sub-soils on wetting. PhD Thesis, Witwatersrand University, Johannesburg, South Africa.

Little, A.L. 1969. The engineering classification of residual tropical soils. *Proc. Spec. Session on Eng. Props. Lateritic Soils. 7th Int. Conf. Soil. Mech. & Found. Eng., Mexico City,* pp. 1-10.

Lumb, P. 1961. The properties of decomposed granite. *Géotechnique* 3: 226-242.

Mason, B. 1949. Oxidation and reduction in geochemistry. *Jour. Geol.* 57(1): 62-72.

Medina, J. 1989. Tropical soils in pavement design. *Proc. 12th Int. Conf. Soil Mech. and Found. Eng., Rio de Janeiro, Brazil* 1: 543-546.

Mitchell, J.K. 1976. *Fundamentals of soil behaviour.* New York: Wiley, pp. 49-57.

Morin, W.J. & Ayetey, J. 1971. Formation and properties of red tropical soils. *Proc. 5th Reg. Conf. for Africa on Soil Mech. & Found. Eng., Luanda, Angola* 1: 45-53.

Pings, W.B. 1968. Bacterial leaching. *Mineral Industries Bulletin,* Colorado School of Mines 2(3): 19 pp.

Ruxton, G.P. & Berry, L. 1957. Weathering of granite and association erosional features in Hong Kong. *Bulletin, Geol. Soc. Amer.* 68: 1263-1292.

Schwartz, K. & Yates, J.R.C. 1980. Engineering properties of aeolian Kalahari sands. *7th Regional Conf. for Africa on Soil Mech. and Found. Eng., Accra, Ghana* 1: 67-74.

Strakhov, N.M. 1967. *The principles of lithogenesis,* Vol. 1. Edinburgh: Oliver and Boyd.

Sueoka, T. 1988. Identification and classification of granitic residual soils using chemical weathering index. *Proc. 2nd Int. Conf. on Geomechanics in Tropical Soils, Singapore* 1: 55-62.

Sweere, G.T.H., Galjaard, P.J. & Tjong Tjin Joe, H. 1988. Engineering properties of laterites in road construction. *Proc. 2nd Int. Conf on Geomechanics in Tropical Soils, Singapore* 1: 421-428.

Tuncer, E.R. & Lohnes, R.A. 1977. An engineering classification for certain basalt-derived lateritic soils. *Engineering Geology* II(4): 319-339.

Uehara, G. 1982. Soil science for the tropics. Engineering and construction in tropical and residual soils. *ASCE Geotech. Div. Spec. Conf., Honolulu, Hawaii*: 13-26.

van der Merwe, D.H. 1965. The soils and their engineering properties of an area between Pretoria North and Brits, Transvaal. DSc Thesis, University of Pretoria.

Vargas, M. & Pichler, E. 1957. Residual soil and rock slides in Santos, Brazil. *Proc. 4th Int. Conf. for Soil Mech. & Found. Eng., London* II: 394-398.

Weinert, H.H. 1964. Basic igneous rocks in road foundations, C.S.I.R. Research Report 218, CSIR, Pretoria, South Africa.

Weinert, H.H. 1974. A climatic index of weathering and its application is road construction. *Géotechnique* 24(4): 475-488.

Wesley, L.D. 1973. Some basic engineering properties of halloysite and allophane clays in Java, Indonesia. *Géotechnique* 23(4): 471-494.

Yamanouchi, T. & Haruyama, M. 1969. *Shear characteristics of such granular soil as shirasu. Memoirs.* Faculty of Eng., Kyushu University. 29(1): 63-64.

Zaruba, O. & Mencl, V. 1976. *Engineering Geology.* Elsevier Scientific, Amsterdam.

CHAPTER 2

Classification of residual soils

L.D. WESLEY
Department of Civil Engineering, University of Auckland, New Zealand

T.Y. IRFAN
Civil Engineering Building, Kowloon, Hong Kong, People's Republic of China

2.1 REASONS FOR A SPECIAL CLASSIFICATION SYSTEM FOR RESIDUAL SOILS

There are specific features or characteristics of residual soils that are not adequately covered by conventional methods of soil classification such as the Unified Soil Classification system. Among these features are the following:

1. The unusual clay mineralogy of some tropical and subtropical soils gives them characteristics that are not compatible with those normally associated with the group to which the soil belongs according to existing systems such as the Unified Soil Classification System.

2. The soil mass in situ may display a sequence of materials ranging from a true soil to a soft rock depending on degree on weathering, which cannot be adequately described using existing systems based on classification of transported soils in temperate climates.

3. Conventional soil classification systems focus primarily on the properties of the soil in its remoulded state; this is often misleading with residual soils, whose properties are likely to be most strongly influenced by in situ structural characteristics inherited from the original rock mass or developed as a consequence of weathering.

The term residual soil needs some clarification as it is not well defined. In any weathering process which converts rock into soil there will be a gradual transition without a clearly defined point at which the material changes from rock to soil. In the profile shown in Figure 2.1, the upper three horizons (IV, V, and VI, the highly, moderately and completely and weathered zones of Fig. 1.8) will tend to behave as soils, while the lower three (I, II and III) will tend to behave more as rocks. Hence the term residual soil should be applied only to the upper three categories (IV, V and VI).

The term saprolite is also used to some extent in describing or classifying residual soils. It is not proposed to give this term a precise definition here, although it is recognised that it generally refers to a residual soil with clear structural features inherited from its parent rock.

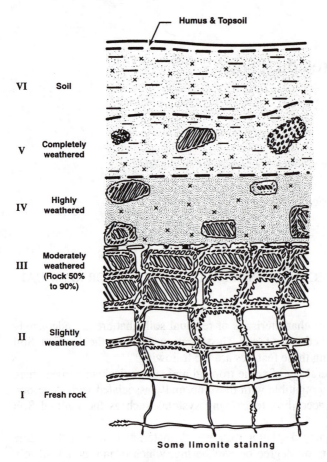

Figure 2.1. Classification of a weathered rock mass profile (after Little 1969).

2.2 CLASSIFICATION SCHEME FOR RESIDUAL SOILS

Wesley (1988) proposed a practical system for classifying all residual soils, based on mineralogical composition and soil micro- and macrostructure. Wesley's classification system is intended to provide an orderly division of residual soils into groups which belong together because of common factors in their formation and/or composition which can be expected to give them similar engineering properties. The system is based on a grouping framework designed to enable engineers to find their way around the rather confused world of residual soils and to enable them to place any particular residual soil into a specific category on the basis of common engineering properties.

This scheme, which is adopted here in a modified form, is not intended to provide an all-embracing method for the detailed systematic description or logging of residual soils in the field, or in the laboratory. It is therefore not intended as a replacement to any particular method of classification at present in use, whether it be a standard method such as the Unified Soil Classification System, or a method proposed specifically for residual soils.

2.2.1 *Basis of the system*

The specific characteristics of residual soils which distinguish them from transported soils can generally be attributed either to the presence of specific clay minerals found only in residual soils, or to particular structural effects, such as the presence of unweathered or partially weathered rock, relict discontinuities and other planes of weakness, and inter-particle bonds. These influences can be grouped under the general headings of composition and structure.

Composition refers to the material of which the soil is made, and includes the particle size, shape, and especially the mineralogical composition of the fine fraction. Composition can be divided into:
– Physical composition, e.g. % of unweathered rock, particle size distribution etc.,
– Mineralogical composition.

Structure refers to the specific characteristics of the soil in its undisturbed in situ state and can be subdivided into two main categories as follows:
   1. *Macro-structure or discernible structure.* This includes all features discernible to the naked eye, such as layering, discontinuities, fissures, pores, presence of unweathered or partially weathered rock and other relict structures inherited from the parent rock mass.
   2. *Micro-structure or non-discernible structure.* This includes fabric, interparticle bonding or cementation, aggregation of particles, pore sizes and shapes, etc.

2.2.2 *The proposed grouping*

The proposed classification is summarized in Table 2.1. The first step in the grouping of residual soils is to divide them into groups on the basis of mineralogical composition alone, without reference to their undisturbed state. The following three groups are suggested:
   1. *Group A*: Soils without a strong mineralogical influence.
   2. *Group B*: Soils with a strong mineralogical influence deriving from clay minerals also commonly found in transported soils.
   3. *Group C*: Soils with a strong mineralogical influence deriving from clay minerals only found in residual soils.

Group A: Residual soils without a strong mineralogical influence

By eliminating those soils that are strongly influenced by particular clay minerals, a group of soils is established which can be expected to have similar engineering properties, or within which further subdivision can sensibly be made to identify such groups.

   In general, soils which have a weathering profile of the type illustrated in Figure 2.1 will come within this group. In relatively rare instances, weathering in the top layer (i.e. zone VI) may be sufficiently advanced for its properties to have become strongly influenced by clay minerals that have developed by extensive weathering.

Group A soils can be further subdivided on the basis of the extent and manner in which their behaviour is influenced by structural effects. It is convenient to separate structural effects into the two broad groups mentioned earlier, namely macro-structure and micro-structure.

Group A can therefore be divided into three main sub-groups:

Sub-group (a)
These are soils in which macro-structure plays an important role in the engineering behaviour of the soil. The lower horizons (IV and V) of the soils which weather according to the pattern shown in Figure 2.1 fall into this category.

Sub-group (b)
These are soils without pronounced macro-structure, but with a strong influence from micro-structure. The most important form of micro-structure is the relict inter-particle bonding or that arising from secondary cementation, and although this cannot be identified by visual inspection it can be inferred from fairly basic aspects of soil behaviour. For example sensitivity is a very good measure of micro-structure, as high sensitivity results from the presence of a distinctive structure (involving some form of bonds) which is destroyed by remoulding. Residual soils which can have a high liquidity index (or can exist in an analogous state) are also those which show pronounced bonding or similar effects which enable the soil to exist in a metastable state close to or above its liquid limit.

Sub-group (c)
Residual soils which are not greatly influenced by macro or micro-structural effects are included here as a third sub-group. However, this sub-group is a very minor group, as very few residual soils of Group A will fall into this category.

Possible further sub-division of Groups A(a) and A (b)
It is recognised that these two groups are rather broad, and possibly too broad to be useful in forming groups on the basis of similar engineering properties. Some further subdivision would appear to be desirable although there is no obvious basis on which to do this. Parent rock is possibly the most influential of several factors which might form a basis for further subdivision, and information on parent rock should therefore be included in classifying and describing these soils.

Group B: Residual soils with a strong mineralogical influence deriving from commonly occurring clay mineral

This group is made up of soils which are strongly influenced by clay minerals commonly found in transported soils. The most significant member of this group is the black cotton soils or 'vertisols', the characteristic properties of which are high shrinkage and swelling potential, high compressibility and low strength. These characteristics are directly related to their predominant mineralogical constituent, namely montmorillonite or similar minerals of the smectite group. The engineering properties of such soils are therefore usually very similar to those of any transported soil consisting predominantly of clay minerals of the smectite group. Structures may have a strong influence on the behaviour of soils in this group, particularly on shear strength and permeability.

Information in the literature suggests that not many other residual soils belong to this group, although there are some residual soils derived from sedimentary rocks which have properties strongly influenced by mineralogical composition.

Group C: Residual soils with a strong mineralogical influence deriving from special clay minerals only found in residual soils

These are the soils that are strongly influenced by the presence of clay minerals not commonly found in transported soils. The two most important minerals involved here are the silicate clay minerals halloysite and allophane. *Halloysite* is a lattice (crystalline) mineral of tubular form and belongs to the same group as kaolinite. *Allophane* is a very distinctive mineral with unusual properties. It is described as amorphous (non-lattice) or gel-like; however recent research suggests it may have a poorly developed crystalline structure. In addition to these silicate minerals, tropical soils may contain non-silicate minerals (or 'oxide' minerals), in particular the hydrated forms of aluminium and iron oxide (the sesquioxides), gibbsite and geothite.

The influence of halloysite and allophane on the soil properties is fairly clear from the case studies recorded in the literature. The influence of the sesquioxides is less well documented. It is convenient therefore to divide this group into three sub-groups:

(a) Halloysitic soils
The principal influence of halloysite appears to be that the engineering properties of the soil are good, despite a high clay fraction, and fairly high values of natural water content in terms of the Atterberg limits (i.e. a high Liquidity Index).

(b) Allophanic soils
Allophanic soils are probably the most distinctive of all residual soils due to the very unusual properties of the amorphous mineral allophane. The influence of allophane is both dramatic and puzzling, in that it results in soils having natural water contents ranging from about 80% to 250%, but which still perform satisfactorily as engineering materials. They are commonly much superior to other soils with water contents of only a fraction of the above values.

(c) Soils influenced by the presence of sesquioxides
The principal role of the sesquioxides appears to be to act as cementing agents which bind the other mineral constituents into clusters or aggregations. With sufficient concentration of sesquioxides, the hard concretionary materials commonly known as laterite are formed. The silica/alumina ratio (SiO_2/Al_2O_3) and the silica/sesquioxide ratio have both been used as indicators of degree of laterization (Morin & Todor 1976). This sub-group could perhaps be termed the lateritic group, but the term laterite is generally used very loosely, sometimes to include both halloysite and allophane and other clay minerals which contain only a trace of sesquioxides, and whose behaviour is not significantly influenced by the sesquioxides.

Table 2.1. Classification of residual soils.

Grouping system		Common pedological names used for group	Descriptive information on in situ state	
Major division	Sub-group		Parent rock	Information on structure
Group A (Soils without a strong mineralogical influence)	(a) Strong macro-structure influence	Give names if appropriate	Give details of type of rock from which the soil has been derived.	Describe nature of structure: – stratification, reflecting parent rock – fractures, fissures, faults etc. – presence of partially weathered rock (state % and physical form, e.g. 50% corestones)
	(b) Strong micro-structure influence	Give names if appropriate		Describe nature of micro-structure or evidence of it: – effect of remoulding, sensitivity – liquidity index or similar index
	(c) Little or no structure influence	Give names if appropriate		Describe evidence for little or no structural influence
Group B (Soils strongly influenced by commonly occurring minerals)	(a) Montmorillonite (Smectite group)	Black cotton soils, Black clays, Tropical black earths, Grumusols, Vertisols		Describe any structural effects which may be present, or other aspects relevant to engineering properties. Evidence of swelling behaviour, extent of surface cracking in dry weather, slickensides below surface etc.
	(b) Other minerals			
Group C (Soils strongly influenced by clay minerals essentially found only in residual soils)	(a) Allophane sub group	Volcanic ash soils, Andosols, Andepts		Give basis for inclusion in this group. Describe any structural influences, either macro-structure or micro-structure.
	(b) Halloysite sub group	Tropical red clays, Latosols, Oxisols, Ferralsols		
	(c) Sesquioxide sub group (gibbsite, geothite, haematite)	Lateritic soils, Laterites, Ferralitic soils, Duricrusts		Give basis for inclusion in this group. Describe structural influences – Especially cementation effects of the sesquioxides.

The above groups, especially the halloysite and allophane groups can be further subdivided on the basis of structure. Allophanic soils (which always appear to be associated with more recent volcanic ash deposits as parent material) show considerable variation in the degree to which structure influences their engineering properties. Some are of quite low sensitivity, while others are of medium to high sensitivity, indicating a strong micro- structural component in their undisturbed state.

The suggested grouping system is shown in Table 2.1. An additional item of information which is usually of major importance in influencing the properties of residual soils is the type of parent rock. This should therefore always be included among the descriptive information regarding any soil type, and in some circumstances can be used as a basis for further subdivision within a particular group.

Table 2.2 lists some of the more distinctive characteristics of these soil groups and indicates the means by which they may possibly be identified.

2.3 PROPERTIES OF THE GROUPS AND METHODS OF IDENTIFICATION

General
A disadvantage of this classification system is its use of mineralogical composition as a starting point for establishing the three principal groups. However this is not an insurmountable difficulty, and can be overcome in several ways, such as the following:

1. In many parts of the world there is a considerable body of knowledge on the occurrence and mineralogy of local soils available to geotechnical engineers to make use of.

2. Most countries have institutions which undertake mineralogical studies (usually for agricultural purposes) which can be made use of by engineers.

3. With some soils, the physical appearance, geological setting, and conventional soil mechanics tests (such as water content and Atterberg limits) are adequate to identify the predominant clay mineral.

The latter point applies particularly to the montmorillonite group and the allophane sub group. Soils of a dark grey or black colour with high liquid limit plotting above the A-line on the Plasticity Chart are very likely to belong to the montmorillonite group. Such soils normally form in low lying poorly drained areas.

Soils formed from volcanic material, especially recent volcanic ash deposits, having high water contents and Atterberg limits, and which undergo pronounced irreversible effects on drying are very likely to belong to the allophane group. Figure 2.2 shows the approximate boundaries of the montmorillonite, halloysite, and allophane groups on the plasticity chart. These boundaries are rather arbitrary and may require amending as further data becomes available. It should be noted that some volcanic ash soils may plot closer to the A-line than shown in Figure 2.2. If this is the case then it may well be that there are clay minerals present other than allophane. The boundaries shown in Figure 2.2 are intended to define groups within which behaviour is likely to be similar due to the predominant influence of the clay mineral indicated.

Properties of the groups are discussed in greater detail below.

Table 2.2. Characteristics of residual soils groups.

Group		Examples	Means of identification	Comments on likely engineering properties and behaviour
Major group	Sub-group			
Group A (Soils without a strong mineralogical influence)	(a) Strong macro-structure influence	High weathered rocks from acidic or intermediate igneous rocks, and sedimentary rocks	Visual inspection	This is a very large group of soils (including the 'saprolites') where behaviour (especially in slopes) is dominated by the influence of the discontinuities, fissures, etc.
	(b) Strong micro-structure influence	Completely weathered rocks formed from igneous and sedimentary rocks	Visual inspection, and evaluation of sensitivity, liquidity index, etc.	These soils are essentially homogeneous and form a tidy group much more amenable to systematic evaluation and analysis than group (a) above. Identification of nature and role of bonding (from relict primary bonds to weak secondary bonds) important to understanding behaviour.
	(c) Little structural influence	Soils formed from very homogeneous rocks	Little or no sensitivity, uniform appearance	This is a relatively minor sub-group. Likely to behave similarly to moderately overconsolidated soils.
Group B (Soils strongly influenced by commonly occurring minerals)	(a) Smectite (montmorillonite group)	Black cotton soils, many soils formed in tropical areas in poorly drained conditions.	Dark colour (grey to black) and high plasticity suggest soils of this group	These are normally problem soils found in flat or low lying areas, of low strength, high compressibility, and high swelling and shrinkage characteristics
	(b) Other minerals			This is likely to be a very minor sub-group.
Group C (Soils strongly influenced by clay minerals essentially found only in residual soils)	(a) Allophane group	Soils weathered from volcanic ash in the wet tropics and in temperate climates	Very high natural water contents, and irreversible changes on drying	These are characterised by very high natural water contents, and high liquid and plastic limits. Engineering properties are generally good, though in some cases high sensitivity could make handling and compaction difficult.
	(b) Halloysite group	Soils largely derived from older volcanic rocks; especially tropical red clays	Reddish colour, well drained topography and volcanic parent rock are useful indicators	These are generally very fine grained soils, of low to medium plasticity, but low activity. Engineering properties generally good. (Note that there is often some overlap between allophane and halloysitic soils).
	(c) Sesquioxide group	This soils group loosely referred to as 'lateritic', or laterite	Granular, or nodular appearance	This is a very wide group, ranging from silty clay to coarse sand and gravel. Behaviour may range from low plasticity to non-plastic gravel.

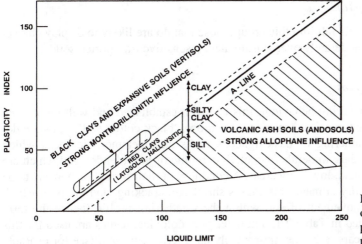

Figure 2.2. Influence of mineralogical composition on position on plasticity chart.

Group A, sub-group (a)

This is a very large group which includes most residual soils derived from the weathering of rocks in well drained situations, i.e. slopes or rises. The behaviour of these soils, especially in slope stability situations is likely to be dominated by the influence of the discontinuities, joints, fissures, and other relict structures. There are three major types of discontinuities in this group; these are:

1. Regular stratification, reflecting that present in parent rocks such as shale or sandstone,

2. Discontinuities, consisting of fractures, fissures, or joints deriving from a variety of causes, such as tectonic or cooling effects,

3. The presence of partially weathered rock, i.e. zones of material intermediate between soil and rock and sometimes referred to as 'saprolite'.

These relict structural features influence the soil behaviour in different ways depending on the engineering situation involved (e.g. Irfan & Woods 1988), but their general effect is to make this group of soils the least amenable to analysis by normal soil mechanics methods.

Group A, sub-group (b)

This group consists of soils where weathering has advanced to the extent that discontinuities and other features of the parent rock are not a significant influence. The most significant property of this group is probably the interparticle bonding which is a common feature of many, if not most residual soils. In the absence of discontinuities this group of soils is amenable to engineering analysis, and studies to investigate the nature and influence of bonding are commonly a valuable approach to understanding their behaviour.

Group A, sub-group (c)

Very few residual soils fall into this group; those that do are likely to display characteristics similar to moderately overconsolidated, insensitive transported soils.

Group B, sub-group (a)

In terms of its engineering behaviour, this is a well established and well recognised group, first identified in soil mechanics literature as 'black cotton soil'. Soils are derived from the weathering of basic igneous rocks and metamorphic rocks which dominantly contain dark minerals (e.g. olivine, biotite or hornblende), and which are found in poorly drained, low lying areas. A condition of their formation appears to be a seasonal wet and dry climate; this causes shrinkage, cracking, and swelling which results in vertical movement of soil within the weathering profile (hence the term vertisol). As indicated in Table 2.1, a number of pedological names are used for this group. Vertisol appears to be the most widely used term at present (see for example Mitchell & Sitar 1982, British Geological Society Engineering Working Party Report 1990).

These soils have the properties typical of any smectite (montmorillonite) rich soil and are very poor engineering materials. They have low strength and high compressibility with high swelling and shrinkage characteristics. Further information on the formation and properties of these soils are given by Lumb (1982) and Horn (1982).

Group C, sub-group (a)

As already stated, this allophane-rich group is probably the most distinctive group of residual soils. Their predominant characteristics include:
 – Very high natural water contents and high values of L.L. and P.L.,
 – Irreversible decrease in plasticity and increase in particle size when air or oven dried,
 – They are likely to have flat compaction curves without distinctive values of optimum water content or maximum dry density,
 – Empirical relationships applicable to transported soils are likely to be misleading when applied to these soils.

Despite these unusual properties, their engineering behaviour is good, especially when the soil is in its undisturbed state (Wesley 1977, Wesley & Matuschka 1988). The soils tend to have high shear strength with moderate to low compressibility, and are remarkably stable on steep natural or cut slopes. However, difficulties are likely to be encountered with earthworks in these soils, firstly because of the wet climates in which they often occur, and secondly because conventional compaction specifications and control methods are not suitable. The soils may in some cases have high sensitivity, making handling and compaction difficult.

The precise nature of the mineral allophane and its extraordinary capacity to retain water is not well understood and further research is necessary to improve knowledge in this area.

Group C, sub-group (b)

The clay mineral halloysite is characterised by very small particle sizes and low activity. However, the precise nature of halloysite does not appear to be well established. Two forms of halloysite have been identified – hydrated halloysite and metahalloysite, (Newill 1961). Both forms appear to consist of tube-shaped particles, although with the dehydrated metahalloysite the tubes may be split or partially unrolled. The most significant property of halloysite soils is that despite small particle size and relatively high plasticity their engineering properties are generally good (Terzaghi 1958, Matyas 1969, Wesley 1974). Various structural concepts have been suggested to account for the good properties of these soils but further research is needed before any definite conclusions can be drawn. Red clays of volcanic parent material are commonly composed predominantly of halloysitic particles and form a predominant member of this group.

Group C, sub-group (c)

This is a very large, rather poorly defined group of soils. The term laterite, or lateritic soils has traditionally been used to designate this group. The intention with the present classification system is to limit this group to those soils in which the concentration of the sesquioxides is sufficiently high to act as a cementing agent to bond particles together in a permanent manner and form concentrations or clusters of particles. There has been a tendency in the past to include in this group any clay found in the tropics having a red colour. Unless there is clear evidence of cementation caused by sesquioxides, red clays should not be included in this group. To be included in this group the cementation must be sufficiently strong that it is not broken down simply by remoulding the soil.

The soils included in this group therefore range from low plasticity silty clays through the concretionary deposits made up predominantly of gravel-sized material. This group could justifiably have a classification system specifically tailored to meet its own needs. Gidigasu's (1976) classification could provide a starting point for development of such a system.

There is generally considered to be a genetic connection between the three Group C soils. With continued weathering allophane is believed to change to halloysite; also the continued weathering of halloysite soils is believed to cause the removal of silica accompanied by the concentration of iron and aluminium oxides to form lateritic materials.

2.4 USE OF THE UNIFIED SOIL CLASSIFICATION SYSTEM IN CONJUNCTION WITH THE PROPOSED SYSTEM

As indicated in the introduction, the justification for using the Unified Soil Classification system (USCS) with residual soils (especially those found in the tropics) has long been questioned, because the engineering characteristics assigned to them by the system do not reflect their true properties. This is particularly the case with soils which plot in the MH zone of the USCS. Many residual soils rich in allophane or kaolin plot in the MH group and according to standard 'guides' used with the USCS these soils have 'poor' engineering properties and are considered unsuitable for various engineering purposes. However, in practice, such soils frequently have good engineering properties. This is especially true of the allophane soils already described.

Although it might be possible to amend the boundaries on the Plasticity Chart and form new zones, it appears that the most practical way to overcome this anomaly is to give the mineralogical composition of the soil together with its classification according to the USCS. If the mineralogical composition is not known, then the origin of the soil or its pedological classification should be given. When this is given then the mineralogical composition may well become evident. Soils of volcanic ash origin with high liquid limit values which plot well below the A-line are almost certain to be rich in allophane.

Giving both the USCS classification and the mineralogical group is in keeping with the intention of the system proposed here. The system is not intended to displace or supplant existing systems, especially the USCS. It is intended to provide a systematic overall grouping and in so doing to complement existing systems.

2.5 EXAMPLE OF CLASSIFICATION OF A RESIDUAL SOIL

As an example, the soil forming the weathered residual andesite profile illustrated in Figure 1.11 will be classified. The mineralogy of the profile is shown in Figure 3.4:

– The clay minerals comprising the soil consist (Fig. 3.4) of kaolinite, muscovite and chlorite, together with quartz. The relative proportions of these minerals vary with depth, but cannot be said to have a strong influence on the properties of the soil. Thus the soil falls within Group A.

– The saprolitic joints in the soil exercise an important influence on the strength of the soil, so it falls into sub-group (a). As the joints in the parent andesite probably resulted mainly from cooling stresses, the discontinuities are of type (b).

Hence the soil classifies as: A (a)(b).

REFERENCES

Geological Society Engineering Working Party Report. Tropical Residual Soils 1990. *Quarterly Journal of Engineering Geology* 23(1): 1-101.

Gidigasu, M.D. 1976. *Lateritic Soil Engineering*. Amsterdam: Elsevier.

Hong Kong Geotechnical Control Office 1988. *Geoguide 3: Guide to rock and soil descriptions.*

Horn, A. 1982. Swell and Creep properties of an African black clay. *Proc. ASCE Specialty Conference on Engineering and Construction in Tropical and Residual Soils, Hawaii*: 199-215.

Irfan, T.Y. & Woods, N.W. 1988. The influence of relict discontinuities on slope stability in saprolitic soils. *Proc. Second Int. Conf. on Geomechanics in Tropical Soils, Singapore* 1: 267-276.

Lumb, W.B. 1982. Engineering problems in tropical and residual soils in Hawaii. *Proc. ASCE Specialty Conference on Engineering and Construction in Tropical and Residual Soils, Hawaii*: 1-12.

Matyas, E.L. 1969. Some engineering properties of Sasumua clay. *Proc. Specialty Session, Seventh International Conference ISSMFE, Engineering Properties of Lateritic Soils, Mexico.*

Mitchell, J.K. & Sitar, N. 1982. Engineering properties of tropical residual soils. *Proc. ASCE Specialty Conference on Engineering and Construction in Tropical and Residual Soils, Hawaii*: 30-57.

Morin, W.J. & Todor, P.C. 1976. Laterite and lateritic soils and other problem soils of the tropics. USAID Publication 3682, Lyon Associates, Baltimore USA.

Newill D. 1961. A laboratory investigation of two red clays from Kenya. *Géotechnique* 11(4): 302-318.

Terzaghi, K. 1958. The design and performance of Sasumua dam. *Proc. Institution of Civil Engineers* 9: 369-394.

Wesley, L.D. 1974. Some basic engineering properties of halloysite and allophane clays in Java, Indonesia, *Géotechnique* 23(4): 471-479.

Wesley, L.D. 1977. Shear strength properties of halloysite and allophane clays in Java, Indonesia. *Géotechnique* 27(2): 125-136.

Wesley, L.D. 1988. Engineering classification of residual soils. *Proc. Second Int. Conf. on Geomechanics in Tropical Soils, Singapore* 1: 77-84.

Wesley, L.D. & Matuschika, T. 1988. Geotechnical engineering in volcanic ash soils, *Proc. Second Int. Conf. on Geomechanics in Tropical Soils, Singapore* 1: 333-340.

CHAPTER 3

Mineralogy and microstructure

J.B. QUEIROZ DE CARVALHO
Atecel Campina Grande, Paraiba, Brazil

J.V. SIMMONS
Sherwood Geotechnical and Research Services, Corinda, Queensland, Australia

3.1 MINERALOGY AND MICROSTRUCTURE RELATED TO WEATHERING

The mineralogy and microstructure of tropical and residual soils are associated with modes of soil formation and occurrence. Microstructure should be considered at two relative scales:
1. Ped and inner-ped voids,
2. Grains and intergranular voids.

Peds are natural aggregations of soil grains which may exist on a scale ranging from centimetres to microns. They are the natural units of mechanical interaction of the soil grains, and may break down or form in response to wetting and drying or various methods of mechanical agitation.

Physical breakdown and chemical reactions are weathering processes leading to the formation of tropical and residual soils. Physical breakdown rates are controlled by exposure and the energy transmitted to the parent material through the local environment. Chemical reaction processes can be summarised as:
– Decomposition,
– Leaching,
– Dehydration.

These processes may proceed simultaneously, cyclically, or sequentially, depending on the climatic conditions and the time of exposure relative to the process reaction rates. Both mineralogy and microstructure are definitively associated with these processes. A wide variety of soils can be produced, depending on the above, and it is necessary to keep a broad perspective when dealing with tropical and residual soils.

Decomposition includes the physical breakdown of the rock fabric and the chemical breakdown of constituent minerals, usually primary rock-forming minerals. Typical products are clay minerals, oxides, hydroxides, and free silica.

Under tropical conditions, reactions may occur relatively quickly so that recently transported soils may subsequently be modified into materials with residual soil characteristics. Reaction rates also vary so that some minerals may have completely

31

decomposed (e.g. feldspars) when neighbouring grains (e.g. quartz) are virtually unaltered.

In contrast, reactions under more temperate conditions occur more slowly so that physical breakdown may dominate the soil formation processes (see also Chapter 1).

Leaching and re-deposition, which include laterization processes, involve removal of combined silica, alkaline earths, and alkalies. There is a consequent accumulation of oxides and hydroxides of sesquioxides. The leached materials may be redeposited and accumulate elsewhere in the soil profile.

Dehydration (either partial or complete) alters the composition and distribution of the sesquioxide-rich materials in a manner which is generally not reversible upon wetting. Dehydration also influences the formative processes of clay minerals. In the case of total dehydration, strongly cemented soils with a unique granular soil structure may be formed.

3.2   MINERALOGY AND OCCURRENCE OF WEATHERING PRODUCTS

Tropical decomposition tends to favour formation of the clay mineral kaolinite. This is by the far most common clay mineral in tropical residual soils. Under suitably moist conditions, halloysites will be formed. Under prolonged decomposition, silica can be removed to the extent that free alumina is present and gibbsite is formed. Usually, part of the silica produced in the soil will be in the form of quartz. Generally, the iron oxides present and/or remaining are sufficient to form goethite and haematite. Iron oxides will form a mineral depending on the in-situ conditions. For example, haematite is only formed in very strong oxidizing conditions, whereas geothite and limonite form where there are conditions of continuous moisture and aeration. Specific minerals are also characteristic of soils from certain parts of the world. For instance, illites are frequently identified in African lateritic soils but not in Brazilian laterites, where normally only kaolinite is present. Montmorillonite or smectite is usually the predominant clay mineral in the very widely occurring black cotton soils, and in many soils residual from sedimentary rocks such as mudrocks and shales.

Other mineralogical components cited in literature and present in tropical and residual soils, may be relatively rare and/or of difficult identification. Such minerals include boehmite, anatase, mixed-layered kaolinite-illite or kaolinite-vermiculite assemblages. There is strong evidence indicating that montmorillonite may be present in lateritic soils, for instance, but only as a very-short term transitory mineral in one part of a weathering sequence.

Tropical weathering of volcanic ashes frequently produces an abundance of allophane, a virtually amorphous clay mineral having an unusually high natural moisture content. Allophane may be identified by its characteristically large, irreversible change of plasticity properties upon drying at different temperatures.

Clay minerals tend to be concentrated in the fine fraction of the soil. Iron oxide minerals such as goethite and/or haematite, and also quartz, tend to be concentrated in the coarse fraction. Gibbsite and boehmite are frequently found as fillings to pores and voids in the concretionary components of the soils. Significant quantities of amorphous components have also been identified in tropical lateritic and saprolitic soils (Queiroz de Carvalho 1981, 1985, 1991).

The significance of mineralogy for the better understanding of these types of soils has been discussed in Chapter 2.

A wide variety of processes may lead to residual soil formation. Highly structured duplex soils tend to form where there is pronounced seasonal wetness and dryness, particularly when the soil profile has matured over a long period of time, as in many parts of Australia (Richards 1985). The reader is also referred to more detailed descriptions (British Geological Society 1991) of the interactions of landform, climate, parent mineralogy, and time.

The classic weathering profile, from mature soil to fresh rock, has been discussed widely, and been subjected to detailed scrutiny in many parts of the World (e.g. Brand 1988). Within the soil layer, there may be differentiation caused by leaching and dehydration. The soil profile is typically structured into an overlying silt (depleted in clays) and a clay-enriched intermediate horizon, often highly fissured in a blocky pattern. It is necessary to maintain a very disciplined profile description methodology (see Chapter 4) in order adequately to record residual soil profiles.

3.3 DETERMINATION OF MINERALOGICAL COMPOSITION

Soil mineralogy can be assessed in various ways. Very specialised techniques have been developed for particular purposes, but the most common approaches are:
 – X-ray diffraction,
 – Thermo-gravimetry,
 – Optical microscopy including polarisation measurements,
 – Scanning or transmission electron microscopy combined with some form of spectral element identification.

Mineralogical identification using these techniques requires specialised training and procedures. Often, combinations of techniques are necessary in order to make definite identifications. The processes are not straightforward because the preparation and measurement process usually alters the minerals. Each technique must therefore be fully understood to be of greatest use.

The X-ray diffraction (XRD) technique is by far the most widely used, but is only appropriate for minerals with distinctive crystallography. XRD can be carried out using oriented or randomly selected soil samples placed in a sample holder with or without resin to fix the samples in place. Specialised techniques are necessary for soils containing significant quantities of iron. Glycolation (replacement of interlayer water by ethylene glycol) causes changes to interlayer spacing and is a means of identification of montmorillonites.

Thermogravimetry (TG) identifies minerals based on changes that occur as dehydration takes place through a range of temperatures. It is generally imprecise except for certain simple minerals which have clear and unambiguous thermogravimetric signatures.

Optical microscopy (OM), used to examine thin sections with crossed polars, is a well established assessment process for primary rock minerals. Sample preparation may involve impregnation with resin, and this may damage the original microstructure to some degree if not done with extreme care. OM can also be applied to some classes of weathered minerals with success. It can be a useful technique for estimating relative abundances of certain minerals, or for assessment of the texture of weathered rocks including fresh and completely weathered grains.

Scanning electron microscopy (SEM) or transmission electron microscopy (TEM) both involve preparation techniques which may displace water and hence damage the original microstructure. Such damage can be minimised with care, and specially designed environmental microscopes can avoid the problem. SEM imaging can be undertaken on lump samples or thin sections. TEM imaging requires dispersed particles or thin sections.

SEM imaging can reveal microstructure details to sub-micron sizes, and is a particularly useful tool for microstructural studies. When combined with some form of microprobe, for example Energy Dispersive Spectroscopy (EDS), measurements can be made of elemental constituents at specific points on the sample surface. From relative abundances of elements, it is then possible to deduce the chemical composition of the material at the point. For many clay minerals, particularly mixed-layered or hydrated forms that are common in residual soils, it is not possible to obtain a conclusive identification unless this process is combined with another technique, for example XRD, which reveals crystal form.

TEM imaging can reveal mineral crystal details at the lattice level, and hence is used to refine knowledge of the clay species. Some form of spectroscopy is normally used to make chemical composition determinations. TEM imaging has been used to visualise the layering within clay minerals, and hence the mineralogical changes which take place during the weathering processes.

3.4 MICROSTRUCTURE OF RESIDUAL SOILS

Soil structure, fabric, and texture are terms referring to the physical arrangement of grains and peds. These arrangements, together with mineralogy, determine engineering behaviour. In conventional soil mechanics, the importance of fabric has been recognised for a long time, and has been well summarised by Rowe (1974). Most aspects of fabric relevant to engineering behaviour are macrostructural (that is, can be observed without magnification).

Microstructure embraces microfabric, composition, and interparticle forces. Investigations of soil microstructure are carried out using optical microscopy, SEM, or TEM. The microstructure of residual soils may reveal cementation and pea contact arrangements, which can lead to a better understanding of the occurrence and engineering performance of such soils.

Collins (1985) extended microfabric studies to residual soils and outlined the models presented in Figures 3.1 and 3.2. These show examples of microfabric organization at three levels:

1. Elementary level,
2. Assemblage level,
3. Composite level.

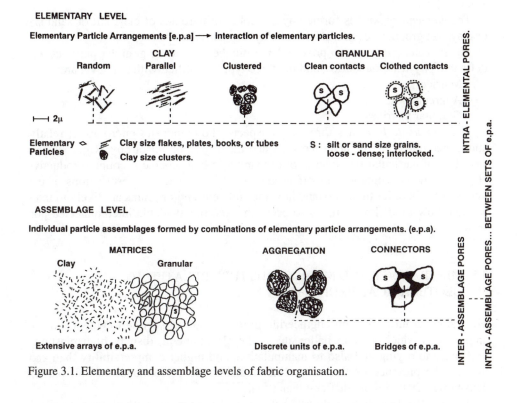

ELEMENTARY LEVEL

Elementary Particle Arrangements [e.p.a] ⟶ Interaction of elementary particles.

CLAY   GRANULAR

Random   Parallel   Clustered   Clean contacts   Clothed contacts

⊢—⊣ 2μ

Elementary ◇ Clay size flakes, plates, books, or tubes
Particles ● Clay size clusters.

S : silt or sand size grains.
loose - dense; interlocked.

ASSEMBLAGE LEVEL

Individual particle assemblages formed by combinations of elementary particle arrangements. (e.p.a).

MATRICES   AGGREGATION   CONNECTORS

Clay   Granular

Extensive arrays of e.p.a.   Discrete units of e.p.a.   Bridges of e.p.a.

INTRA - ELEMENTAL PORES.

INTRA - ASSEMBLAGE PORES... BETWEEN SETS OF e.p.a.

INTER - ASSEMBLAGE PORES

INTRA - ASSEMBLAGE PORES... BETWEEN SETS OF e.p.a.

Figure 3.1. Elementary and assemblage levels of fabric organisation.

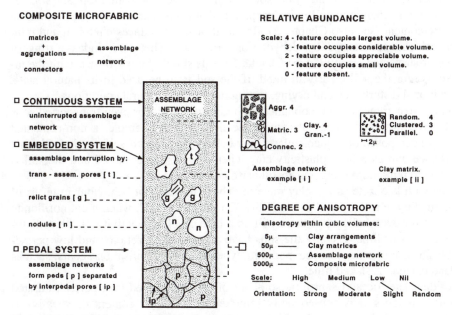

COMPOSITE MICROFABRIC

matrices
+
aggregations ⟶ assemblage
+   network
connectors

RELATIVE ABUNDANCE

Scale: 4 - feature occupies largest volume.
        3 - feature occupies considerable volume.
        2 - feature occupies appreciable volume.
        1 - feature occupies small volume.
        0 - feature absent.

□ CONTINUOUS SYSTEM ⟶ ASSEMBLAGE NETWORK

uninterrupted assemblage
network

Aggr. 4
Matric. 3   Clay. 4
           Gran.-1
Connec. 2

Random. 4
Clustered. 3
Parallel. 0
⊢—⊣ 2μ

□ EMBEDDED SYSTEM

assemblage interruption by:

trans - assem. pores [t]

relict grains [g]

nodules [n]

Assemblage network
example [i]

Clay matrix.
example [ii]

DEGREE OF ANISOTROPY

anisotropy within cubic volumes:

5μ ——— Clay arrangements
50μ ——— Clay matrices
500μ ——— Assemblage network
5000μ ——— Composite microfabric

□ PEDAL SYSTEM ⟶

assemblage networks
form peds [p] separated
by interpedal pores [ip]

Scale: High   Medium   Low   Nil
Orientation: Strong   Moderate   Slight   Random

Figure 3.2. Composite level of fabric organisation.

The *elementary level* is formed by a number of particles of clay, silt or sand size or clay size group, or even a cluster of clay size particles.

The *assemblage level* comprises a large number of clay or granular particles with definable physical boundaries, and the three types identified with this level are:
– Matrices,
– Aggregations,
– Granular matrices.

The *composite level* has three main aspects: the composite microfabric, relative abundance reflecting the heterogeneity, and the degree of anisotropy. Rigorous classification of particular soil microfabrics using a formal scheme is counterproductive. However, the concepts are important and should be related to observations of particular soils, in order to understand how the soil peds and/or grains are likely to interact, and how fluid flow is likely to occur through the available interconnected pore space.

3.5 MINERALOGY AND MICROSTRUCTURE RELATED TO GEOTECHNICAL PROPERTIES

Mineralogical influences on engineering properties are usually self-evident. For example, soils which contain halloysite as the clay mineral, display high plasticity, sensitivity to drying and also to manipulation, and higher compressibility than kaolinites. The presence of this clay mineral makes the performance of soil particle density (G_s) testing more difficult than usual.

A distinctive feature of halloysite soils, even more notable with allophane soils, is the irreversible change in soil index properties that accompanies drying from natural moisture content (see also Chapter 5). Removal of water induces chemical and mineralogical changes caused by dehydration processes. Therefore, index testing of known or suspected Group C (see Chapter 2) soils should be performed with the intended use for the soil clearly in mind. If the soil is to be used in its natural state, with minimal disturbance and drying, testing must reflect natural conditions.

Similar experiences can be obtained with Group A soils, or even transported soils which have sustained some subsequent weathering. For example, a non-cemented aged sand composed of volcanic clasts may behave as a sand if subjected to minimal disturbance, but as a high plasticity clay when the clasts are altered and the soil is subjected to remoulding (e.g. cut-to-fill earthworks).

Soils with kaolinite as the clay mineral exhibit normal test behaviour in terms of low to medium plasticity and permeability. On the other hand, when montmorillonite is present (or even halloysite) there will be a high level of reactivity of the soils towards cement or lime. Usually the soils have a very high plasticity and low permeability. These properties may be modified by addition of stabilisation agents, such as portland cement or slaked lime.

Residual lateritic and saprolitic soils are often encountered in a partly saturated condition. If the soil is subsequently saturated or its water content drastically increased, the final shear strength will depend upon the type of clay mineral present. If kaolinite is the dominant clay mineral, the effect of wetting may not be great. However the presence of halloysite or montmorillonite may result in a large decrease in the shear strength when the soil is wetted.

The influence of sesquioxides is related to their mode of occurrence in the soil. If they cement the particles, they tend to cause a reduction in plasticity. However, if there is only very weak cementation, remoulding of the soil can lead to an increase in the interactions between clay particles, and therefore to an increase in plasticity. Also, the presence of goethite/haematite as discrete particles in the matrix soil, can be enhanced by the process of stabilization, forming cementitious compounds and thereby increasing the strength of the soil. Collins (1985) attributed increased stiffness in a saprolite to the bridging or bracing between quartz grains by clay minerals and/or sesquioxides. This may also result in an increased cohesional strength for the soil.

Baynes & Dearman (1978) concluded that changes of the geotechnical properties of weathered granites were due to microfracturing and intergranular porosity, observed in the microstructure of the weathered material studied. Collins (1985) found similar features on weathered granite from Brazil. Queiroz de Carvalho (1991) found that a highly-cemented microstructure was present in soils which were also highly reactive to the process of lime stabilisation. The voids in these soils were filled by a form of amorphous aluminous material.

Dispersive soil behaviour is often identified in soils which are subjected to cyclic wetting and drying under conditions where clay minerals become sodium-dominated. Simmons (1989) found that microstructural aggregation of clays was the most identifiable factor in occurrence of soils subject to severe dispersive erosion. These clay minerals appeared to aggregate in specific zones thought to be associated with the evaporation zone of seasonal groundwater changes. The study suggested that microfabric, rather than clay chemistry, had the controlling influence. The erodibility of a soil was found to be greatest where silt-sized aggregations were present.

In summary, residual soils must be expected to vary very widely in their engineering properties. Where specific microstructural or mineralogical features are identified, corresponding influences on engineering behaviour may be expected. The most important principle is that residual soils may have much better engineering characteristics than index tests suggest, particularly where index tests are correlated with the engineering behaviour of transported soils as in the USCS system. The geotechnical specialist must be aware of this when assessing a soil for any purpose. Soil properties must be assessed in regard to the particular purpose, or range of purposes, expected. Where there is any doubt about the likely behaviour of such soils, field-based assessments and/or tests for recognition of special behaviour are of much greater significance than for soil mechanics practice for transported soils.

3.6 EXAMPLE OF THE MINERALOGY OF A RESIDUAL PROFILE

As an example, the weathering and resulting mineralogy of the residual andesite soil described in Section 1.5 will be explored (Blight 1996).

Minerals are not equally susceptible to chemical breakdown and mineralogic change as a result of weathering. For each rock there is a sequence of weathering that depends on the minerals present. The first minerals to crystallize out from a magma have the highest internal energy associated with them, and therefore are the most unstable and will break down first during the weathering process. Conversely, the last minerals to crystallise out will have the lowest internal energy and will be the most resistant to weathering. This point is illustrated by Figure 3.3 which shows the crystallization and weathering orders of minerals in andesite according to Bowen (1928) and Arnold, (1984). It will be seen that the two schemes are closely related.

Figure 3.4 shows the distribution of minerals observed in a profile of residual andesite from under the 'Above Africa' surface near Johannesburg, South Africa (Brummer 1980). It will be seen that the andesite weathers to form muscovite, chlorite, kaolin and quartz, although montmorillonite has also been reported to occur (Falla 1985). The analysis shown in Figure 3.4 does not include the haematite or limonite that gives the soil its characteristic colour. It will be noted that quartz occurs in only part of the profile, although where it does occur, it amounts, at depths of between 5 and 10 m, to almost 50% of the soil. These high quartz contents probably did not form part of the original mineralogy of the andesite, but resulted from the secondary deposition of quartz in joints in the andesite. It is very common to find quartz veins in residual andesite profiles. Note that one does not usually find a regular progression of minerals with depth from those characteristic of a high degree of weathering to those characteristic of lesser weathering. This is because the profile is composed of a number of successive lava flows, each of which originally had a slightly different mineralogy and each of which was exposed to weathering for a different time before being covered by the next lava flow.

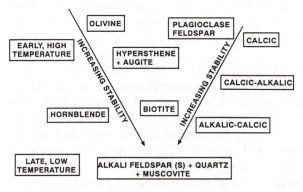

crytallization of minerals in andesite (Bowen, 1928).

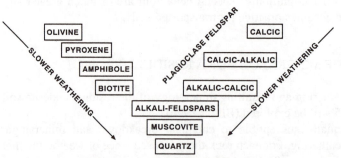

weathering of minerals in andesite (Arnold, 1984).

Figure 3.3. Crystallization and weathering orders of minerals in andesite.

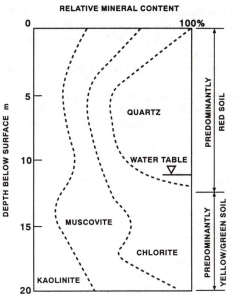

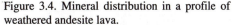

Figure 3.4. Mineral distribution in a profile of weathered andesite lava.

REFERENCES

Arnold, M. 1984. The genesis, mineralogy and identification of expansive soils. *Proc. 5th Int. Conf on Expansive Soils, Adelaide, Australia*: 32-36.

Baynes, F.J. & Dearman, W.R. 1978. The microfabric of a chemically weathered granite. *Bulletin of the International Association of Engineering Geology* 18: 91-100.

Blight, G.E. 1996. Properties of a soil residual from andesite lava, *Proc. 4th Int. Conf. on Tropical Soils, Kuala Lumpur, Malaysia*: 575-580.

Bowen, N.L. 1928. *The evolution of the igneous rocks*. Dover, New York, USA.

Brand, E.W. 1988. Evolution of a classification scheme for weathered rock. Written discussion, Geomechanics in Tropical Soils. *Proceedings of the 2nd International Conference on Geomechanics in Tropical Soils, Singapore* 2: 515-518. Rotterdam: Balkema.

British Geological Society Engineering Group 1991. Tropical Residual Soils. Working Party Report. *Quarterly Journal of Engineering Geology*: 1-101.

Brummer, R.K. 1980. The engineering properties of deep highly weathered residual soil profiles. MSc (Eng) dissertation. Witwatersrand University, Johannesburg, South Africa.

Collins, K. 1985. Towards characterization of tropical soil microstructure. *Proceedings of the 1st International Conference on Geomechanics in Tropical Lateritic and Saprolitic Soils, Brasilia* 1: 85-96.

Falla, W.J. 1985. On the significance of climate and weathering in predicting geotechnical characteristics of residual soils developed on igneous rocks. PhD Thesis, Witwatersrand University, Johannesburg, South Africa.

Queiroz de Carvalho, J.B. 1981. Study of the microstructure of lateritic soils using Scanning Electron Microscope. *Proceedings of the International Seminar on Lateritization Processes, Sao Paulo* I: 563-568.

Queiroz de Carvalho, J.B. 1985. Effects of deferation and removal of amorphous silica and alumina on selected properties of Brazilian lateritic soils. *Proceedings of the International Seminar on Laterite, Tokyo* II: 559-567.

Queiroz de Carvalho, J.B. 1991. Microstructure of concretionary lateritic soils. *Proceedings of the IX Panamerican Conference of Soil Mechanics and Foundation Engineering* I: 117-128.

Richards, B.G. 1985. Residual Soils of Australia. In Brand E W & H B Phillipson (eds), *Sampling and Testing of Residual Soils*. Hong Kong: Scorpion Press, pp. 23-30.

Rowe, P.W. 1974. The importance of soil fabric and its relevance in engineering practice. Rankine Lecture, *Géotechnique* 24(3): 265-310.

Simmons, J.V. 1989. Preliminary Studies of Dispersive Soil Fabrics from the Burdekin River Irrigation Area using the Scanning Electron Microscope. Research Bulletin Number CS38, Department of Civil and Systems Engineering, James Cook University of North Queensland, Townsville, Australia, 43pp.

CHAPTER 4

Profile description and sampling methods

J.V. SIMMONS
Sherwood Geotechnical and Research Services, Corinda, Queensland, Australia

G.E. BLIGHT
Civil Engineering Department, Witwatersrand University, Johannesburg, South Africa

4.1 PRINCIPLES AND PURPOSES OF PROFILE DESCRIPTION

The objective of profile description is to convey information relevant to engineering intentions. The information required is project-dependent, and is always subjected to compromises which include cost and risk.

An examination for the purpose of profile description has to include some form of sampling of the materials. Sampling may range from visual assessment of exposures in excavations, to sophisticated extraction of 'undisturbed' samples. In situ evaluation of properties is preferable to sampling in many circumstances, and is discussed in Chapters 7, 8, 9 and 10. Sampling also involves compromises between cost and the need to reduce design or construction risks through the gathering of information.

The following aspects of purpose and end-use should be considered when planning the choice of profile description and sampling methods:
1. Use for the information,
2. Key principles of material behaviour,
3. Methods for assessment,
4. Methods for sampling,
5. Available background information.

4.1.1 *Use for the information*

The intended end-use usually controls the extent and detail of the description and sampling activities. Reconnaissance studies, where an overview is required early in the activity for a project, can be based very much on appearance, judgement, and experience, since there is usually an opportunity to return for more specific information at a later stage. Detailed design requires site-specific information, such as basic classifications and also basic design parameters. Tender and contract documentation should include sufficient information to give a fair assessment of site conditions, construction issues, and risks for all parties involved. Construction control and quality assurance is dependent upon the provision of adequate and practical information on soil profiles.

41

4.1.2 *Key principles*

The difference between geological and geotechnical information must be recognised. For geotechnical purposes, the material behaviour in the appropriate engineering context must be quantified, whereas geological information is usually qualitative by nature. While qualitative information is important for geotechnical purposes, it is usually not sufficient for design.

Various geotechnical classification schemes have been developed to predict material behaviour semi-empirically. Chapter 2 includes a classification of particular importance to assess the engineering behaviour of residual soils. As pointed out in Chapter 2, a general classification scheme such as the Unified Soil Classification System (USCS) may be used provided that its shortcomings in relation to residual soils are recognized.

Profile description information should include a description of the soil in its natural state (water content, colour, structure, texture, consistency, interactions with water) as well as a recognition of the effects of construction (i.e. use of the soil as a construction material for founding, or as a cut or natural slope, or as a compacted fill).

4.1.3 *Methods of assessment*

The assessment must recognise the limitations and biases imposed by the method(s) available for the profile investigation. These may include:
  – Field inspection of natural exposures,
  – Inspections of test pits and auger holes,
  – Examination of drillhole cuttings,
  – Drillhole cores,
  – In situ probing, vane shear, pressuremeter, etc.
Usually some combination of the above, supplemented by other in situ and/or laboratory testing, is required for an adequate assessment of the soil profile and its properties. In transported soil profiles it may be sufficient to explore indirectly by means, for example, of wash boring, standard penetration testing or piezocone probing. In residual soil profiles, it is almost essential to explore the soil profile visually and tactilely via test holes. This has long been standard practice in certain countries (e.g. Jennings et al. 1973) and has more recently been recommended as an international practice (Cook & Newill 1988).

In countries where water tables are deep it is a common practice to explore the profile by descending a 750 mm to 1000 mm diameter augered test hole in a bosun's chair, or by means of a chain ladder. Most residual profiles are cohesive, and provided one does not descend below the water table and that the holes are profiled within a few hours of drilling, the danger of collapse of the hole is remote. The danger of collapse can also be overcome by temporarily casing the hole with a protective skeleton cage made of steel hoops and longitudinals. Holes should also be tested for the presence of gas, either methane or carbon dioxide, before descending. The soil in its pristine condition is then examined by cutting away the drilling smear on the side of the hole with a geological pick and inspecting, testing or sampling the soil thus exposed in the side of the hole.

4.1.4 *Methods of sampling*

Except for material which is already at the natural ground surface, all sampling involves some degree of disturbance which cannot be avoided. Sampling methods must be chosen with care, in the context of the project requirements. The choice of appropriate sampling techniques usually has to be finalised in the field. Preparation for fieldwork should include allowances for all likely sampling activities, and the best use must be made of available equipment.

Although the work is old, the monumental US Waterways Experiment Station report (Hvorslev 1948) remains as valid today as it was when published 50 years ago.

4.1.5 *Background information*

The maximum value can be obtained from fieldwork only when all available background information is gathered and analysed. The analysis should include an evaluation of the reliability of the information. Planning of all aspects of the fieldwork is another necessary component which should be included. The following are suggestions for steps in planning fieldwork:

– Make a preliminary desk study using available topographic, geological or soil maps, reports, surface and aerial photographs. Consult the records of local library or institutional resources such as municipalities, government departments, research groups, published literature, etc.

– Assess possible equipment requirements for fieldwork, machinery and technology requirements for investigations, including sampling, transportation, etc.

4.2 PROFILE DESCRIPTION

Because all sampling involves some degree of disturbance, it is necessary to distinguish facts related to the soil, from facts related to the method of inspection or sampling. The engineering properties of the material in its natural field state may be difficult to assess, except by including observation of the behaviour of excavation equipment, drills, or probes during the investigation.

Often, for legal purposes or because clients believe that they are getting more cost-effective information, factual reporting only is required. In practice, however, some degree of interpretation is essential. Whatever the case, reporting should always clearly distinguish between factual information and interpreted information. For example, classification information is always based on interpretation. It is good practice to indicate interpretation clearly on logging sheets, either by use of parentheses or by clearly marked sections for 'interpretation'.

4.2.1 *Profile description procedures*

Procedures for profile description have been developed by a wide variety of organisations, and are set out in procedure manuals or codes of practice. Particular procedures may be adapted or extended for other purposes. There is a temptation to include detail for its own sake rather than for specific purposes. This detail may, however, become valuable later on, for reasons not earlier envisaged.

The following procedure list is not expected to be entirely complete, but it should serve as a basis for most purposes. It may be supplemented or reduced according to requirements. The appropriate sections of selected codes of practice may also be substituted. The final choice for use of any such lists lies with the person responsible for the fieldwork.

Because residual soil profiles may grade into weathered and fresh rock, it is necessary to include descriptions appropriate to both soil and rock. It is important to include descriptions of weathering grade for rocks, as assessed in the field. However, interpretations of weathering profiles for classification purposes should be clearly separated from factual descriptions.

4.2.2 *Site records*

The following should be recorded:
– General description of site, general location, vegetation, access,
– Time and date,
– Weather,
– Precise location details (co-ordinates, marks, reference features),
– All field activities (diary, logging forms, equipment used).
Descriptions should be recorded for the following aspects of the soil:
– Moisture condition (M),
– Colour (best based on standard colour charts) (C),
– Consistency (C),
– Soil (e.g. clay, silt, sand, gravel, sandy silt, silty clay, etc.) (S),
– Structure or fabric (zoning, fissuring slickensiding, cementing, quartz veins, etc.) (S).
– Origin (rock type, eg. residual from basalt, residual from shale) (O).

The mnemonic MCCSSO is useful to remember when recording data for a soil profile. Field logs will include information from fieldwork alone. Therefore, descriptions of plasticity, moisture condition, etc. will be qualitative field descriptions. Subsequent laboratory testing may cause modification of the plasticity description, and allow inclusion of numerical moisture content values.

An example of a profile description is given in Figure 1.11. Other examples are given in Figures 4.1a and 4.1b. A description format is given in the appendix. This is based on the code of practice for site investigations used in Australia, but shares most of the common features that are included in the best international practice.

4.2.3 *Recording information on or with logging sheets*

The following guidelines are recommended:
– Use a standardised format,
– Record confirmed boundaries with full lines, inferred boundaries by dashed lines,

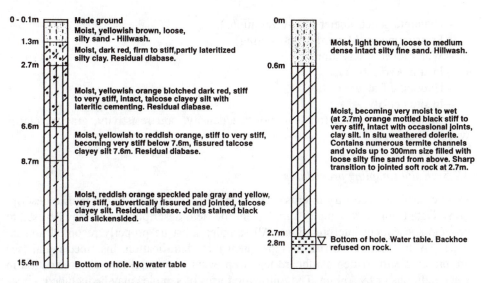

Figure 4.1. a) Profile of residual diabase recording laterization of upper part of profile and saprolitic jointing, and b) Profile of residual dolerite recording sharp transition to rock and results of termite activity.

– Enter data on a field copy at the time of assessment, and transfer to a clean copy immediately after each day's work,

– Append records of probing as appropriate,

– Take photographs wherever possible (camera should have close-up lens and flash),

– Make sketches of the site to clarify layouts and features, either for subsequent drafting or for later informal briefing,

– Sketch and dimensions positions of each trench, hole, or natural exposure that is logged,

– Include mapping dimensions or references, and cross-references to fieldbooks including those of other parties, e.g. survey.

4.2.4 *Insitu testing records*

It is preferable to undertake some form of field testing, even if using only simple hand equipment. For larger-scale and more complex investigations, in situ testing may be more sophisticated. All in situ testing must be recorded or cross-referenced on logging sheets. In situ tests may include:

– Hand-held shear vane,

– Hand-held penetrometer,

– Light probing (e.g. dynamic cone penetrometer),

– In situ vane shear test,

– Standard penetration test (its is essential to observe best practice regarding water level control, rod dimensions, and hammer mass and free fall),

– Static cone penetrometer,

– Dilatometer,

– Dynamic penetrometer (many variations),
– Pressuremeter (Menard or self-boring),
– Water level observations,
– In situ permeability,
– Hydraulic fracture test,
– In situ density,
– Geophysical probes (nuclear moisture/density, soil resistivity, gravity, elastic wave velocity, etc.).

4.2.5 *In situ sampling procedures*

In all cases it is necessary to cross-reference all such samples to the profile descriptions. Data from these samples supplement the field descriptions and may be used to provide subsequent interpretations. All samples must be properly recorded and labelled so that there is no risk of subsequent mis-identification. Inscribed metal tags are preferable to written cardboard tags to prevent accidental loss of information in wet conditions or by soiling. The following forms of samples may be included:
– Disturbed bag samples or bulk samples,
– SPT samples,
– Thin-walled tube or open-drive tube samples,
– Thin-walled tube piston samples,
– Mutli-barrel core samples,
– Block samples.

4.2.6 *Core logging procedures*

Good quality core, and good quality core logging, requires careful attention to drilling conditions, and use of appropriate equipment. The following aspects need to be recorded:
a) Details of coring equipment,
b) Details of rate of advance, flushing, loss of water, drill response etc.,
c) Orientation of core (a range of techniques, great care required),
d) Use of a reference line (to orient all core segments in the core tray prior to logging),
e) Identification of geological discontinuities (separate from fractures or breaks induced by drilling),
f) Noting of all features having geotechnical implications (e.g. core losses may be more significant than the recovered core).

4.2.7 *Hand samples/visual estimation*

It is important to describe what is experienced with hand samples. For plasticity, an estimation of 'low' or 'high' is all that can be reasonably expected. For moisture condition, all that can be reasonably described is the degree to which free water is apparent. This can be recorded as dry, damp, moist, very moist, or wet. Simple hand balling tests for estimating moisture relative to Plastic Limit or Optimum Moisture

Content are also useful. The important factors to be included in the field description are:
- Appearance,
- Soil behaviour type (indicated as an interpretation),
- Field or hand test results.

Many residual soils have an appearance and engineering characteristics in the undisturbed state that are very different to those in the remoulded state. Field descriptions may thus not correspond to classifications based on laboratory testing (see Chapter 2). It is important that the field personnel be involved in the assessment of final descriptions which include laboratory test data.

Where there are apparent conflicts, information should not be discarded or hidden. The preferred solution is to clearly mark on the logs the field description, and include the laboratory-based description or classification in parentheses.

4.2.8 *Logging of equipment performance*

Excavation and handling characteristics of materials are of critical importance for construction activities. Standardised geotechnical logging procedures should therefore record all relevant details of the equipment used for the field investigations, including
- Weight of excavator, bucket size and capacity, power rating,
- Weight of drilling rig,
- Type of drive (top, bottom, or kelly bar),
- Rod size and lengths,
- Flushing medium (air, foam, mist, water, mud, additives etc.).

Although qualitative, some estimate of performance of the equipment will also assist. For example, drilling or coring penetration rate ranging from 'fast' to 'slow' can be logged based on timing or on the drillers' assessment. Technological advances in drilling include automatic logging of machine characteristics such as rotation rate, torque and thrust, and penetration rate. Such information may be included on logs, but appropriate background explanations must also be provided.

4.3 SAMPLING METHODS

4.3.1 *Purpose of samples*

In reality there are no truly undisturbed samples, because stress changes and some mechanical disturbance are unavoidable during sampling. Disturbance is therefore assessed in terms of influence on the engineering properties which are to be measured from the sample.

The intended purpose of the sampling must be clearly understood when the fieldwork is carried out. Field personnel should take all possible precautions to preserve samples in as close to the undisturbed state as possible.

For soil tests which involve remoulding material or changing moisture conditions from the field condition, there may be no purpose in trying to preserve any of the fabric or intact soil features. Bulk samples, collected in impervious, sealed bags, are suitable, but small representative moisture content samples must also be included. It is important not to let the soil samples dry out prior to testing. (See Chapter 5 for the effect of drying on soil properties of residual soils). At the other end of the scale, to estimate in situ properties from samples requires great care during sampling, preservation, transportation, and storage. High quality piston, thin-walled tube, or core samples have to be handled in a very delicate manner, to preserve the material in as close to the undisturbed condition as possible.

For all rock core, no matter what the lithology, moisture content should be preserved by wrapping in several thickness of plastic film (cling wrap). The only exception is where there is no intention to test the core other than by immediate field techniques, and where it is judged that moisture change will not lead to core degradation that would change the geotechnical assessment of the rock mass. This latter point is very important since core represents three things: intact rock material, natural rock defects, and sampling-induced defects or material disturbance.

Brand & Phillipson (1985) discuss details of practices from many parts of the world, for sampling and testing of tropical residual soils. Perhaps the greatest difficulty lies with high-quality sampling and testing of weathered profiles that include soil and rock in various conditions and consistencies. No single approach to such a situation is possible. The best advice under such conditions is to utilise personnel with the best possible training and greatest possible experience, and to undertake all work with the purpose of the sampling clearly in mind.

The above assumes that the laboratory which receives the samples has the capabilities to undertake testing of an appropriate quality. There must be a careful matching of testing competence and capabilities to the efforts involved in collecting samples, or the requirement to collect samples at all. Poor quality information can be as misleading as wrong information. The quality of the fieldwork must therefore be determined with a realistic appreciation of all the steps involved in obtaining test data.

Where sample quality and disturbance become major issues, consideration must be given to the deployment of in situ testing procedures. The costs and benefits of in situ testing versus sampling requirements should be carefully evaluated as part of the planning phase for the fieldwork.

4.3.2 *Sampling methods*

The methods introduced in Section 4.2.6 are elaborated below. This is done as a checklist of points which require particular consideration with residual soils, in addition to the normal practices associated with each method. Figures 4.2, 4.3 and 4.4 illustrate sampling devices and procedures suitable for sampling residual soils.

a) *Disturbed samples or bulk samples*:
– Ensure adequate quantity for purpose,
– Maintain separate small carefully sealed representative subsamples for moisture condition,
– Take care with sealing and handling to avoid sample damage or loss of moisture,
– Measure moisture condition as soon as possible, particularly if bulk sample is stored for more than a day under uncontrolled conditions. Moisture content should preferably be measured on site.

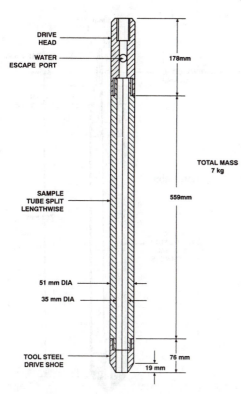

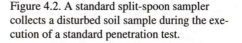

Figure 4.2. A standard split-spoon sampler collects a disturbed soil sample during the execution of a standard penetration test.

b) *SPT samples*:

– Take note of moisture with sample, and record whether moisture condition is a reliable measure of in situ condition,

– Ensure sample is adequately sealed and labelled,

– SPT samples should be retained for inspection during tendering, where applicable.

c) *Thin-walled or open-drive tube samples*:

– Tubes must clean,

– Cutting edges must be in good condition, and area ratio measured and recorded,

– Samples should be obtained by hydraulic thrust. Hammering cannot be allowed,

– Actual depth range sampled, and sample recovery, must be recorded,

– Conduct hand-tests where possible, immediately upon sampling,

– Seal samples immediately: non-shrink wax or a wax-petroleum jelly mix is preferred to o-ring plugs, but second-stage sealing in slightly moistened plastic bags is always advisable,

– Store with in situ axis horizontal in stable temperature conditions (shaded, humid)

– Transport gently, packed in shock-absorbent material such as saw-dust or polystyrene beads and handle as fragile goods,

– Labelling should include hole number and location date, depth to top of sample, length of sample and also identify the top of the sample.

d) *Piston samples*:

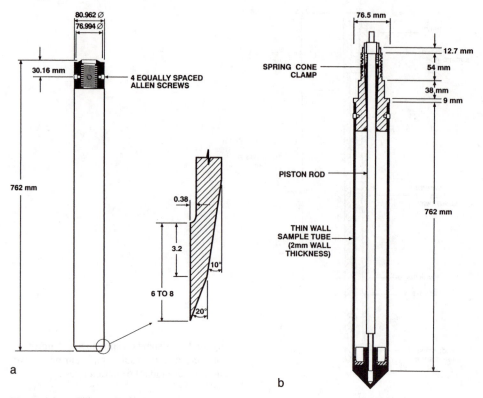

Figure 4.3. a) Thin-walled tube sampler used to sample firm to stiff soils, b) Thin-walled stationary piston sampler, suitable for very soft soils.

– Comments for thin-walled or open-drive samples apply here as well,

– Handle very gently, seal immediately after measurement of sample length recovery, with every precaution to avoid moisture loss especially from sands and silts,

– Samples should be tested on site if at all possible, as transport may inflict too much disturbance.

e) *Core samples*:

– Coring equipment operator should be highly trained and closely supervised,

– Core runs should be minimised in broken ground or highly weathered rock, to avoid losses and minimise disturbance,

– Essential features only should be logged initially, followed immediately by photography and sealing to minimise moisture loss,

– Store in shaded and well ventilated area, if possible where moisture condition can be controlled.

f) *Block samples*:

– Have all equipment ready prior to starting sampling process,

– Work as quickly as possible, recording all in situ features on each face of the block,

– Seal progressively,

– Storage and transportation to be as for tube, piston, or core samples to minimize disturbance,

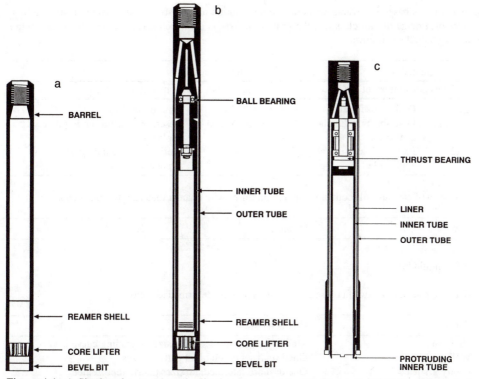

Figure 4.4. a) Single tube core-barrel effective in rock sampling but unsuitable for undisturbed sampling of soils and soft rocks, b) Double tube core-barrel and the triple tube core-barrel, c) used to sample soft rocks

– Test as soon as possible, to minimise effects of storage.

The appendix gives examples of soil logging sheets that can be used to formalize the recording of soil and rock profiles.

APPENDIX – PROFILE DESCRIPTION AND SAMPLING METHODS

Descriptive terms – Soil

Soils may be described for logging purposes as follows.
   – Unified classification (in brackets if not based on laboratory tests):
   – Overall soil name: Plasticity or particle size, colour,   colour is preferably described
   – Secondary component(s): Plasticity or particle size, colour,   with reference to a standard
   – Minor component(s): Plasticity or particle size, colour,   colour chart.
   – Moisture condition: Consistency or density; structure; additional observations.
   The terms used in the description are defined below.

Particle size distribution

	.002		0.06	0.2		0.6	2	6	20	60 mm		
	Fine	Medium	Coarse	Fine	Medium	Coarse	Fine	Medium	Coarse	Cobbles	Boulders
Clay		Silt			Sand			Gravel				

Soil name is based on overall particle size distribution and plasticity. As most soils are combinations of a range of particle sizes, the primary matrix is described and modified by minor components generally as follows:

Coarse grained soils		Fine grained soils	
% Fines	Component	% Coarse	Component
≤ 5	Omit or use 'trace'	≤ 15	Omit or use 'trace'
> 5 < 12	Describe as 'with clay/silt' as applicable	> 15 < 30	Describe as 'with sand/gravel' as applicable
> 12	Prefix soil as 'silty/clayey' as applicable	> 30	Prefix soil as 'sandy/gravely' as applicable

Plasticity of clay and silt both alone and in mixtures with coarser material, is described as:

Low	Medium	High
LL = 35%	LL = 50%	

(LL = liquid limit)

Grading characteristics of coarse grained soils may be described as follows:

Term	Characteristics
Uniform	Majority of grains of similar size, ie grading curve is very steep
Well graded	Good representation of all particle sizes
Poorly graded	One or more intermediate sizes poorly represented
Gap graded	Deficient in certain intermediate sizes

Colour is described for soil in the 'moist' 'in profile' condition

Moisture condition

'Dry'	Looks and feels dry. Cohesive soils hard and brittle. Granular soils free-running.
'Moist'	Feels cool, darkened in colour. Cohesive soils can be moulded by hand. Granular soils tend to cohere.
'Wet'	Feels cool, darkened in colour. Cohesive soils easily moulded by hand, granular soils tend to cohere. Free water on hands when remoulding.

Consistency of cohesive soils is based on the undrained shear strength and is generally estimated with or without the aid of pocket penetrometer or hand vane.

Term	Undrained shear strength kPa	Field guide to consistency
Very soft	≤ 10	Exudes between the fingers when squeezed in hand
Soft	> 10 ≤ 20	Can be moulded by light finger pressure
Firm	> 20 ≤ 50	Can be moulded by strong finger pressure
Stiff	> 50 ≤ 100	Cannot be moulded by fingers. Can be indented by thumb
Very stiff	>100 ≤ 200	Can be indented by thumb nail
Hard	>200	Can be indented with difficulty by thumb nail

Density of non cohesive soils

Term	SPT N Value
Very loose	< 4
Loose	4-10
Medium dense	10-30
Dense	30-50
Very dense	> 50

Descriptive terms – Rock

Rocks are described using the terms defined below in the following order:
- Rock name, grain size, texture, colour, (composition),
- Strength, weathering, (condition of rock material),
- Structure, defects, rock mass weathering, (rock mass properties).

Grain size

Average grain size	Descriptive term
> 2 mm	Coarse
> 0.06 < 2 mm	Medium
> 0.002 < 0.06 mm	Fine
< 0.002 mm	Amorphous or cryptocrystalline

Texture refers to individual grains. Typical terms include porphyritic, crystalline, granular, oolitic, schistose etc.

Colour: Described in moist condition using simple terms modified as necessary by 'light', 'dark', 'mottled'. Borderline conditions are described by combination (e.g. red-brown, not reddish-brown). Colour is preferably described with reference to a standard colour chart.

Strength: strength refers to the strength of the rock material, not the rock mass. The following terms, based on Point Load Strength Index, are used.

Term	Letter symbol	Point Load Strength Index, (MPa)	Field guide to strength
Extremely low	EL	≤ 0.03	Easily remoulded by hand to a material with soil properties
Very low	VL	> 0.03 ≤ 0.1	Material crumbles under firm blows with sharp end of geological pick. Can be peeled with knife: too hard to cut triaxial sample by hand. Pieces up to 3 cm thick can be broken by fingers.
Low	L	> 0.1 ≤ 0.3	Easily scored with knife. Indentations 1 mm to 3 mm show in the specimen with firm blows of pick point. Has dull sound under hammer. A piece of core 150 mm long. 50 mm diameter may be broken by hand. Sharp edges of core may be friable and break during handling.
Medium	M	> 0.3 ≤ 1.0	Readily scored with a knife: a piece of core 150 mm long, 50 mm diameter can be broken by hand with difficulty.
High	H	> 1 ≤ 3	A piece of core 150 mm long, 50 mm diameter cannot be broken by hand but can be broken by a geological pick with a single firm blow: rock rings under hammer.

Term	Letter symbol	Point Load Strength Index, (MPa)	Field guide to strength
Very high	VH	> 3 ≤ 10	Hand specimen breaks with geological pick after more than one blow: rock rings under hammer.
Extremely high	EH	>10	Specimen requires many blows with geological pick to break through intact material: rock rings under hammer.

Weathering

Term	Symbol	Field identification
Fresh	F or Fr	Rock substance unaffected by weathering.
Slightly weathered	SW	Rock substance affected by weathering to the extent that partial staining or partial discolouration of the rock substance (usually by limonite) has taken place. The colour and texture of the fresh rock is recognisable: strength properties are essentially those of the fresh rock substance.
Moderately weathered	MW	Rock substance affected by weathering to the extent that staining extends throughout the whole of the rock substance and the original colour of the fresh rock is no longer recognisable.
Highly weathered	HW	Rock substance affected by weathering to the extent that limonite staining or leaching affects the whole of the rock substance and signs of chemical or physical decomposition of the individual minerals are usually evident. Porosity and strength may be increased or decreased when compared with the fresh rock substance usually as a result of the leaching or deposition of iron. The colour of the original fresh rock substance is no longer recognisable.
Extremely weathered	EW	Rock substance affected by weathering to the extent that the rock exhibits soil properties i.e. it can be remoulded and can be classified according to the USCS, but the texture of the original rock is still evident.

Structure: Refers to the large scale inter-relationship of textural features in the rock mass. Typical common terms include:
– Bedded, laminated, massive (sedimentary rocks),
– Foliated, banded (metamorphic rocks),
– Massive, flow banded (igneous rocks)

Where appropriate, spacing or thickness of structural features are given (in mm usually). Unquantified descriptive terms (widely spaced, thinly bedded etc) are not used.

Defects are fractures in the rock mass and include joints, faults, shear planes, cleavages, bedding partings etc. Details of all relevant defects are given, using such terms as open, tight, infilled, plane, curved, slickensided, etc.

Thickness, openness, spacing etc are given in millimetres: unquantified descriptive terms are not used.

RQD (Rock Quality Designation): The ratio of length of core recovered in pieces of 100mm or longer to the length of core run drilled, expressed as a percentage.

Rock mass weathering: For large scale exposures (e.g. cut rock slopes), but not for boreholes, the weathering of the rock mass may be described in the following terms:

Grade	Descriptive terms
1A	Fresh: no visible signs of rock material weathering
1B	Fresh except for limonite staining or major defect surfaces
II	Some to all of the rock mass is discoloured by slight weathering
III	Less than half the rock mass is moderately to extremely weathered. Fresh or slightly weathered rock is present either as a discontinuous framework or as corestones.
IV	More than half the rock material is moderately to extremely weathered. Fresh or slightly weathered rock is present either as a discontinuous framework or as corestones
V	The rock material is extremely weathered with the original mass structure still largely intact
VI	Refer to soil classification system

REFERENCES

Brand, E.W. & Phillipson, H.B. (eds) 1985. *Sampling and testing of residual soils.* Hong Kong: Scorpion Press.

Cook, J.R. & Newill, D. 1988. The field description and identification of tropical residual soils. *2nd Int. Conf. on Geomechanics in Tropical Soils, Singapore* 1: 3-10.

Hvorslev, M.J. 1948. *Subsurface exploration and sampling of soils for civil engineering purposes.* Waterways Experiment Station, Vicksburg, Miss., USA.

Jennings, J.E., Brink, A.B.A. & Williams, A.A.B. 1973. Revised guide to soil profiling for civil engineering purposes in Southern Africa. *The Civil Engineering in South Africa*, Jan, pp. 3-12.

CHAPTER 5

Classification and index tests

A.B. FOURIE
Civil Engineering Department, Witwatersrand University, Johannesburg, South Africa

It is well-known that oven-drying, and even air-drying, affects the properties of soils, although this effect is usually small for transported soils. Because of their origin, by slow in situ decomposition in a largely anaerobic environment, residual soils are particularly prone to changes in properties caused by drying and exposure to air. Drying can cause partial or complete dehydration of the clay minerals and can change them and their properties irreversibly. Even air-drying at ambient temperature can cause changes that cannot be reversed by re-wetting, even if the re-wetted soil is allowed to mature for long periods.

Apart from the relatively well-known effect of drying on index properties of residual soils, drying also affects the composition, compressibility and shear strength characteristics of residual soils (e.g. Frost 1976).

5.1 MOISTURE CONTENT

The conventional test for the determination of moisture content is based on the loss of water when a soil is dried to a constant mass at a temperature between 105 and110°C. In many residual soils, however, some moisture exists as water of crystallisation, within the structure of the minerals present in the solid particles. Some of this moisture may be removed by drying at the above temperature, i.e. not only free water is driven off. This is illustrated by Figure 5.1, which shows the effects of different drying temperatures compared with standard tests dried at 105°C. As shown, the moisture contents may apparently increase significantly with drying temperature. Figure 5.2 shows how the apparent value of the water content increases progressively for four residual clays as the temperature of drying increases from 20°C to 40°C (and relative humidities of 30%) to the standard drying temperature of 105°C. (The effect is even more pronounced for soils containing halloysite or allophane (Terzaghi 1958)). The option of air-drying soils is problematic as this may take an extremely long time in a humid environment. The following procedure is therefore recommended:

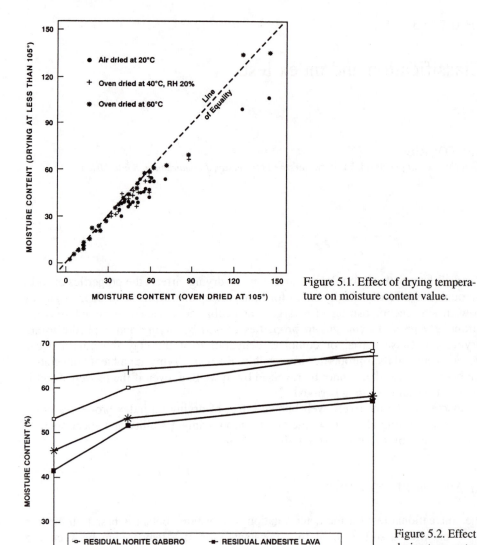

Figure 5.1. Effect of drying temperature on moisture content value.

Figure 5.2. Effect of
drying temperature
on apparent mois-
ture content of four
residual clays.

Two test specimens should be prepared for moisture content determinations. One
specimen should be oven dried at 105°C until successive weighings show no further
loss of mass. The moisture content should then be calculated in the normal way. The
second sample should be air-dried (if feasible), or oven-dried at a temperature of no
more than 50°C and a maximum relative humidity (RH) of 30% until successive
weighings show no further loss of mass. The two moisture content results should
then be compared; a significant difference (4-6% of moisture content obtained by
oven-drying at 105°C) indicates that 'structural' water is present. This water forms
part of the soil solids, and should therefore be excluded from the calculation of

moisture content. If a difference is detected using the two different drying procedures, all subsequent tests for moisture content determination (including those associated with Atterberg limit tests, etc.) should be carried out by drying at the lower
temperature (i.e. either air-drying, or oven-drying at 50°C and 30% relative humidity). If possible, the lower drying temperature of 50°C should be used.

5.2 ATTERBERG LIMITS

In addition to the problem of moisture content determination discussed above, two
further problems may be experienced when carrying out tests to determine the Atterberg limits, optimum moisture content and maximum dry density of residual soils.

5.2.1 *The effect of pre-test drying*

The effect of air-drying specimens prior to carrying out the Atterberg limit tests,
rather than testing, starting at natural moisture content has been observed to result in
a decrease in the liquid limit and plasticity index, (Terzaghi 1958, Rouse et al. 1986,
Wesley & Matuschika 1988). This phenomenon is illustrated in Figure 5.3.

According to Townsend (1985), the effect of drying prior to testing may be attributed to:

a) Increased cementation due to oxidation of the iron and aluminium sesquioxides,
or

b) Dehydration of allophane and halloysite, or

c) Both a) and b) above.

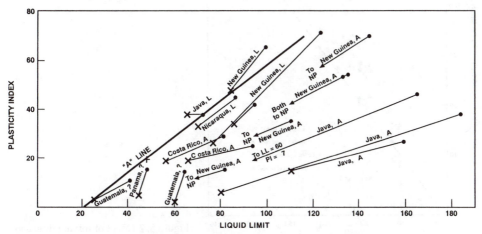

Figure 5.3. Effect of drying on plasticity of volcanic soils (Morin & Todor 1975).

In order to be meaningful, Atterberg limit tests should therefore be performed without any form of drying prior to carrying out the test. If some form of drying is unavoidable, for example, if the water content exceeds the plastic limit, this should be noted on the laboratory report, and details given of the method and duration of drying.

Usually, Atterberg limit tests are carried out on that part of the soil finer than 425 µm. The above recommendation usually precludes separating this fraction. As far as possible, all large soil particles should be removed by working the undried soil with the fingers through a 2.38 mm opening sieve.

5.2.2 *The effect of duration and method of mixing*

In general, the greater the duration of mixing (i.e. the greater the energy applied to the soil prior to testing), the larger the resulting liquid limit, and to a lesser extent, the larger the plasticity index. This has been attributed to longer mixing resulting in more extensive breaking down of cemented bonds between clay clusters and within peds, and thus the formation of greater proportions of fine particles. This effect is illustrated in Figure 5.4. An extreme example of the effect of time of mixing on Atterberg limits is illustrated by Figure 5.5 which shows how the number of blows to close the groove in the Liquid Limit cup increased from 20 after minimal mixing to 235 after 25 minutes of mixing. The soil in this case was a residual mud rock.

In order to address this problem, the following procedure is recommended:

Five test specimens should be mixed with water to give a range of moisture contents suitable for liquid and plastic limit determinations. The minimum amount of air drying should be used, and preferably none at all. This should not be too difficult as

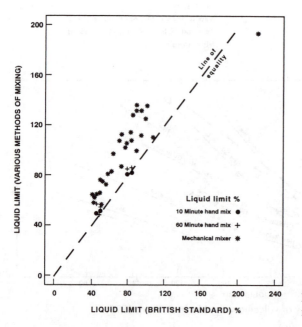

Figure 5.4. Effect of mixing time and means on liquid limit.

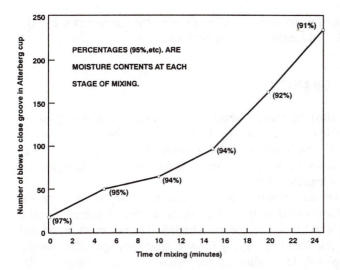

Figure 5.5. Effect of time of mixing on number of blows to close groove in liquid limit cup.

the in-situ moisture content of the majority of soils is at or below the relevant plastic limit. The mixing time should be standardized at 5 minutes, and the mixed specimens should be left for moisture content equilibration overnight before testing.

On the following day the liquid limit should be determined using a standard procedure (e.g. British Standard 1377, Part 2) with a minimum of further mixing. A subsample from each of the specimens used in the test should be used for determining the moisture content, using the procedure established after conducting the evaluation described in Section 5.1. The remainder of each specimen should then be mixed continuously for a further 25 minutes before again determining the liquid limit. A significant difference (i.e. > 5% of liquid limit obtained from test a on a specimen mixed for 5 minutes) between the liquid limits from tests using 5 and 30 minute mixing times indicates a disaggregation of the clay sized particles in the soil. If this disaggregation is confirmed by a repeat of the above procedure, the entire programme of testing should:

1. Limit the mixing times to no more than 5 minutes.

2. Make use of fresh soil for each moisture content point in Atterberg limit tests, as well as in compaction tests.

The soil should be broken down by soaking in distilled water, and not by drying and grinding. The soil should be immersed in distilled water to form a slurry, which is then washed through a 425 μm sieve until the water runs clear. The material passing the sieve is collected and used for the Atterberg limit tests. The particles retained on the sieve should be dried and weighed. The percentage by dry mass passing the 425 μm sieve should be calculated and recorded.

In cases where the soil contains appreciable quantities of soluble salts, distilled water should not be used as it may cause a change in the properties of the soil. The soluble salt content of the soil is most conveniently assessed by measuring the conductivity of a soil paste at the liquid limit. If the conductivity exceeds $50 mSm^{-1}$, distilled water should not be used for mixing. In this case an equivalent 'soil water' can be prepared by making up a 2:1 by mass mixture of distilled water and soil (about a 5:1 by volume mix of water to soil). After allowing the mixture to settle, the clear water is decanted and used for the Atterberg limit tests.

5.3 PARTICLE SPECIFIC GRAVITY

The specific gravity (also called the particle density) (G_s) is used to calculate parameters such as clay fraction, void ratio and porosity. In residual soils the specific gravity may be unusually high or unusually low. It is thus essential that the specific gravity be determined in the laboratory using an accepted standard test procedure, for the purpose of carrying out these calculations.

The soil to be used in this test should be at its natural moisture content. Pre-test drying of the soil should be avoided as this tends to reduce the measured specific gravity as compared with natural moisture content samples. The dry mass of the soil used in the test should be calculated by drying the soil specimen *after* the specific gravity test has been completed. Depending on the outcome of the evaluation described in Section 5.1, it may be necessary to air-dry the soil, or to dry at a reduced temperature.

5.4 PARTICLE SIZE DISTRIBUTION

As with the above parameters, the particle size distribution of a residual soil may be affected by certain aspects of sample preparation, as described below.

1. *Effect of drying*: The most widely reported effect of drying is to reduce the percentage that is reported as the clay fraction (finer than 2 µm). It is accordingly recommended that drying of the soil prior to testing be avoided. The soil sample should be split into two sub-samples: one for determining the moisture content (in order to calculate initial dry mass), and the other for the particle size distribution test. The latter sample should be immersed in a solution of dispersant such as dilute alkaline sodium hexametaphosphate, and thereafter washed through the standard nest of sieves.

2. *Chemical pre-treatment*: This should be avoided wherever possible. Pre-treatment with hydrogen peroxide is only necessary when organic matter is present. If it is considered necessary to eliminate carbonates or sesquioxides, then pre-treatment with hydrochloric acid is used.

3. *Sedimentation*: It is essential to achieve complete dispersion of fine particles prior to carrying out a sedimentation test. The use of alkaline sodium hexametaphosphate is suggested. In some instances a concentration of twice the standard value may be required. If the above dispersant is ineffective, an alternative such as trisodium phosphate should be used. *In all cases the dispersant solution should be freshly made before use in the laboratory.*

REFERENCES

Frost, R.J. 1976. Importance of correct pretesting preparation of some tropical soils. *Proc. 1st S.E. Asian Conf. on Soil Engineering, Bangkok, Thailand.*

Rouse, W.C., Reading, A.J. & Walsh, R.P.D. 1986. Volcanic soil properties in Dominica, West Indies. *Engineering Geology* 23: 1-28.

Terzaghi, K. 1958. Design and performance of Sasumua Dam. *Proc. I.C.E.* 9: 369-394.

Townsend, F.C. 1985. Geotechnical characteristics of residual soils. *Journal of Geotechnical Engineering, ASCE* 111(1): 77-94.

Wesley, L.D. & Matuschika, T. 1988. Geotechnical engineering in volcanic ash soils. *Proc. 2nd Int. Conf. Geomechanics in Tropical Soils, Singapore* 1: 333-342.

CHAPTER 6

Compaction

J.V. SIMMONS
Sherwood Geotechnical and Research Services, Corinda, Qld., Australia

G.E. BLIGHT
Civil Engineering Department, Witwatersrand University, Johannesburg, South Africa

6.1 PURPOSES AND PROCESSES OF COMPACTION

Compaction is the process of densifying soil and reducing air voids by applying mechanical energy. Densification leads to improvement in the engineering properties of strength, compressibility and stiffness, and to a reduction in permeability.

There are some basic scientific principles involved in the compaction process. However, the means of achieving the desired degree of compaction involves a combination of technology and judgement, and the recognition that soil materials are inherently variable. Efficient compaction is an art that is dependent on engineering skill and judgement. Compaction is undertaken in the field using the best available technology. The compacted product requires compromises between energy and cost expended, and the value of the result obtained. Engineering design must recognise the reality of what can be achieved in the field.

Residual soils, are widely used as construction materials, mainly as fill for embankment dams and road embankments and as selected layers in highways and airfield construction. Certain residual soils, such as those containing smectite or halloysite clays may be unsuitable for these uses, either because of inadequate strength, or excessive change of volume with varying water content, or because of loss of strength on wetting. However, smectitic and halloysitic materials have been used to form impervious layers in water-retaining embankments. Examples of such uses are Sasumua dam described by Terzaghi (1958) and the Arenal dam described by Rodda et al. (1982).

The parent rock is usually variable in composition, particularly if it is an igneous rock. The degree of weathering will also be variable. Hence the selection of representative samples for testing can be a major problem. For the same reason good control of quality in compacted fills of residual material may be extremely difficult to achieve.

For example, Figure 6.1 (Blight 1989) shows the variability with depth of the grading analysis and Atterberg limits in a profile of residual weathered norite gabbro. Note the sharp transition from a clay to a sand at a depth of 4 m. There is usually a

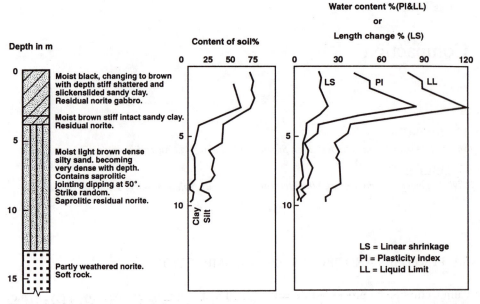

Figure 6.1. Variation in vertical direction of soil composition and index properties for a weathered norite gabbro profile.

similar variation in lateral extent. This variability makes selection of consistent material or consistent blending of material very difficult to achieve in practice.

Drying of the soil from its in situ water content may change both its index and compaction properties (see Chapter 5). Hence soil samples have to be treated and tested with the greatest care if the results of compaction tests are to be at all meaningful. The influence of sample preparation and laboratory procedure on the compaction characteristics of a lateritic soil (Gidigasu 1974) is well illustrated by Figure 6.2. Not only was the optimum water content of the soil significantly altered by air- or oven-drying before compaction, but the maximum dry density was also considerably changed.

The compaction characteristics of residual soils may be very dependent on the method of applying the compactive energy. In particular, laboratory compaction curves may bear little resemblance to the compaction curve achievable in the field. This phenomenon (which may also apply to transported soils) is illustrated by Figure 6.3a which compares roller and laboratory compaction curves for a weathered granite pegmatite (Blight 1962). In one case it was not possible to achieve laboratory maximum dry density in the field because the optimum water content for roller compaction was more than 3% lower than that for laboratory compaction. In the second case (6.3b) it was not possible to achieve laboratory maximum dry density with roller compaction. However, the comparison is better than it appears from the figure, as maximum dry density for roller compaction actually represents 98% of the laboratory maximum. Figure 6.3c shows the grading curves for the two materials. This data also illustrates the variability of material from a single borrow pit. Material A was obtained by blending the upper, more weathered layers of soil, while Material B was a blend of the less weathered underlying soil.

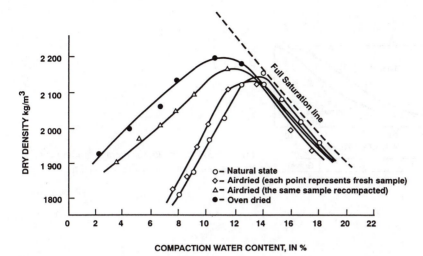

Figure 6.2. The influence of method of sample preparation and laboratory procedure on compaction characteristics of lateric soil from Ghana.

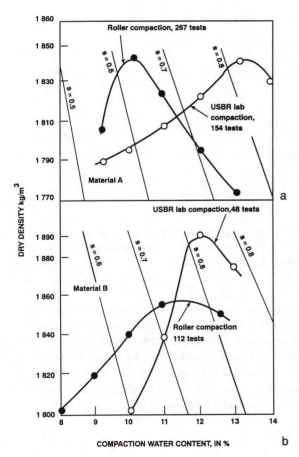

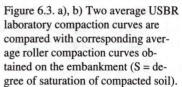

Figure 6.3. a), b) Two average USBR laboratory compaction curves are compared with corresponding average roller compaction curves obtained on the embankment (S = degree of saturation of compacted soil).

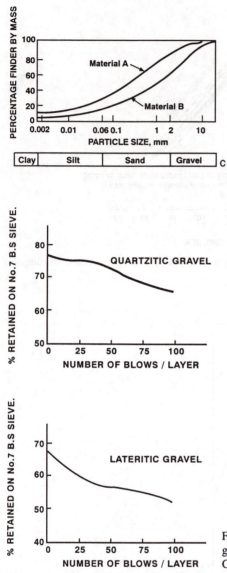

Figure 6.3. Continued. c) Average grading curves for materials A and B of Fig 6.3a and b.

Figure 6.4. Effect of increasing compaction on gravel size content of two residual granite soils from Ghana.

Compaction often results in progressive break-down of the particles in a residual soil. For example, Figure 6.4 (Gidigasu & Dogbey (1980) illustrates the progressive breakdown of particle size under compaction of a quartzitic gravel and a lateritic gravel both residual from the weathering of granite. In cases like this, it becomes imperative to use a fresh soil sample to establish each point on the compaction curve, otherwise the compaction characteristics of the soil will change progressively as the compaction test proceeds, and the test result may be meaningless.

6.1.1 *Special nature of residual soils*

Some special characteristics of residual soils must be understood clearly, if the compaction process is to be understood and the effort and cost of compaction is to be optimised. Residual soils include soils with all of the characteristics of the transported soils, discussed widely in geotechnical literature. The following characteristics, while not a complete list, are associated with special considerations for efficient compaction of residual soils. Residual soils, especially those of volcanic and igneous origin, often have:

- High in situ moisture contents,
- Metastable clay minerals,
- Soil structures that are lightly cemented,
- Weathered soil particles that break down under compactive effort,
- Sesquioxide minerals that are affected by wetting and drying.

The starting point for any engineering use of soil is the in situ, borrow-pit or 'as won' condition. Many tropical and subtropical environments are characterised by frequent or seasonal heavy rainfall. Handling of soils under these conditions may be difficult and the characteristics of the soils themselves may add further complexity to the problem of effective compaction.

6.1.2 *Compaction mechanisms*

Compaction may be undertaken on soils in situ, or soils that have been excavated, transported, and placed. The structure and physical nature of the soil may have a major influence on the choice of compaction method.

Free water (that is, water within the soil mass that is not chemically bound by the clay minerals) can have a controlling influence on the ease with which a given degree of compaction is achieved. Within certain bounds, this free water reduces soil strength, assisting break-down of the soil structure to reduce the air voids. As the free water content increases, there is a stage beyond which the compaction-facilitating effect is offset by the increased energy required to move the air out of the soil. Once the air permeability of the densifying soil reaches the stage where the air cannot escape, void reduction is impeded. The presence of water then becomes a factor that prevents further compaction from taking place. This simple observation leads to the concept of an optimum moisture condition for a given compaction process. The optimum condition is a function of the soil physical properties, initial structure, and the means by which mechanical energy is imparted to the soil mass.

The optimum moisture condition for a given compaction procedure can be illustrated roughly as follows, using the degree of saturation of the soil (S) to distinguish different moisture content states:

$S < 15\%$: Water reduces soil strength, air void space is freely interconnected and rapid expulsion of air and densification can occur,

$15\% < S < 40\%$: Reduction of soil strength is offset by reducing air permeability of soil. This is the optimum range for compaction,

$S > 40\%$: Reduction of air permeability becomes more significant than reduction of soil strength. Air voids become discontinuous and further densification cannot take place.

The degree of saturation is however, not a convenient measure for compaction control. The objective of compaction is to achieve maximum densification appropriate to the engineering purpose. Compaction is therefore usually measured in terms of dry density achieved. The optimum moisture content is associated with the achievement of a maximum dry density for a given compactive effort. Usually this occurs at an air voids content (A_v) of between 5% and 3%. Moisture contents in excess of the optimum are associated with air voids of less than 3% which remain almost constant as water content increases. At this stage additional water adds to void space and leads to a direct reduction in the dry density.

Figure 6.5 illustrates compaction curves for a wide range of residual soils subjected to the same method of compaction and compactive effort. All curves have an optimum for compaction. Very low densities and abnormally high in situ moisture contents may be associated with volcanic soils.

All soils exhibit the following traits:

– High strengths and void contents in the moisture content regime dry of optimum,

– A rapid reduction of strength in the moisture content regime wet of optimum, where the curve follows a line of minimum air voids,

– A range of moisture contents around the optimum condition, where the acceptability of the compacted soil can be measured by a variety of methods. In situ strength is often the most appropriate compaction control parameter for residual soils.

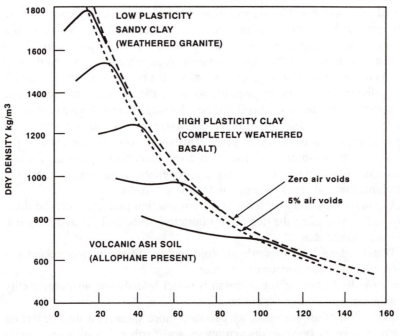

Figure 6.5 Typical compaction curves for residual soils showing optimum characteristic.

6.1.3 *Field compaction process*

The choice of compaction equipment should be made with a view to minimising earthworks costs while achieving the desired engineering properties for the compacted soil. Earthworks design must include a consideration of what range of equipment is available, and any practical constraints such as weather and site conditions which may influence what can be achieved in the field.

The optimum moisture condition for field compaction is best determined for any compaction equipment by a process of field trials. It is recommended practice to include field trials as an initial component of the construction program. The trials can be used to optimise equipment selection and operation, and to identify the field optimum moisture condition (see Fig. 6.3). Where the soil being compacted is sufficiently uniform in its physical properties, field trials can also be used to finalise the method of compaction, with construction being controlled by this means.

Another factor to be considered is the process of moisture conditioning. It is relatively easy to add water to the soil, but more difficult to ensure that the added water is uniformly distributed within the soil. In high rainfall areas, it may be necessary to dry the soil in order to operate compaction equipment. Air drying may be slow, and must be aided by ploughing or tilling to turn the wetter soil to the surface and expose it to sun and wind.

There are a number of detrimental effects on the engineering properties of the compacted soil, if effective and uniform mixing of water and soil is not achieved. These include:
 – Dry clods of soil in a wet matrix, resulting in large voids in the compacted mass,
 – Shearing due to distortion of over-wet fill under the action of the compaction plant, resulting in loss of shear strength of the compacted mass,
 – De-bonding between compacted layers, resulting in loss of shear strength in a horizontal direction. This can be avoided by scarification immediately prior to addition of loose fill layers. De-bonding of layers also has a detrimental effect on permeability,
 – Poor trafficability of construction plant, and/or ponding of water: this can be avoided by proper attention to grades and levels during construction.

The final product should have the desired engineering properties. It is therefore important that adequate supervision and quality control are used during the fill construction process.

6.1.4 *Consequences of unsatisfactory compaction*

It can be very difficult and expensive to rectify inadequate compaction. Because of the time that may be involved in obtaining test results, there will be a tendency for substandard work to be buried prior to being identified. It is therefore important to understand some of the serious consequences of inadequate compaction:
 – Rutting, cracking, and excessive settlement of embankments causing loss of ride quality or dangerous unevenness,
 – Slumping of embankment slopes, and/or loss of freeboard in water storages, possibly causing overtopping, breaching and disastrous flooding,

– Leakage and/or piping erosion leading to overall embankment failure, followed by flooding or loss of water,
   – Inadequate performance results in a loss of utility, and also to excessive capital and/or maintenance costs for rectification,
   – In situations where there are no resources for remedial works, the entire project may fail once the intended utility is lost.

6.2 FUNCTION AND RESPONSIBILITY FOR COMPACTION

The intention should be to achieve adequate performance from the compacted materials to meet the design requirements. There are particular responsibilities for all participants in the process but especially the geotechnical specialist and the constructor. This can be understood in terms of the function of the compacted fill.

6.2.1 *Function of compacted fill*

Compacted fill acts as a structure or a structural platform, supporting some form of structure which may have a variety of performance requirements. Alternatively, fill may be a construction platform, in order to achieve site trafficability, support for formwork, or a stable soil raft on which a structure can be supported.
   Compacted fill may also be used as a component of a retaining structure. The retained material is often water, but other functions include retaining dams for slurries or other fluids (tailings, leachates, hydrocarbons) or for solids (slump debris, landfill wastes). Specialised performance criteria apply to most retaining structures.

6.2.2 *Role of the geotechnical specialist*

The geotechnical specialist is usually involved in investigation, and may have a lesser role in detailed design, documentation, and construction. The assessment of borrow sources for use as fill materials may involve an investigation for design purposes.
   The design report and specification development for construction purposes may be undertaken by others, but should include specialist guidance. Ideally, specialist geotechnical involvement in assessment of fill compaction works is required, as part of site and/or quality control supervision.
   Other aspects requiring specialist assessment are:
   – Disputed construction,
   – Supervision of construction of earthworks,
   – Failures of earthworks,
   – Suitability and reliability of construction performance testing.

6.2.3 *Role of the constructor*

All construction with earth materials involves an element of risk due to unforeseen variability. The constructor is required to assess the:
   – Suitability and availability of construction plant,

– Productivity and costing for compaction activities,

– Responses to weather changes, material source changes, and other events that may arise related to fill placement and compaction,

– Suitability and reliability of performance testing.

In some forms of contract, the constructor is responsible for setting and maintaining performance testing activities. The risk to the end-user is that the activities will be undertaken in such a manner as to excessively favour the constructor's interests. The end-user also may not have the technical skills to understand whether the required end-product is being achieved.

In other methods of contract, the end-user or agency sets and maintains performance testing activities. This can result in the constructor being held accountable for end-product failures which were really a function of inadequate design, specification, supervision, and testing.

6.3 COMPACTION CONTROL

There are many reference texts on geotechnical testing and on construction quality control. The intention of this section is to highlight two aspects of compaction control in the form of checklists: criteria for control, and methods for achieving control.

6.3.1 *Criteria for compaction control*

In order to set appropriate criteria, the function of the fill must be understood, and a number of additional constraints must also be considered:

– Construction requirements related to weather and climate (limited time for completion, rain, wind, low or high temperatures, etc.),

– Moisture conditioning requirements for source material,

– Availability and standards of resources for quality control,

– If available, performance data from field compaction trials either for method specification (uniform source material) or optimum compaction characteristics for equipment used,

– Availability and standards of any reference laboratory testing on which specifications have been based.

Table 6.1 is a summary of tests which have been used for compaction control. These can be used as a guide to selection of appropriate testing. The frequency of testing should be related to a number of additional factors, and must be determined separately.

6.3.2 *Methods for compaction control*

Table 6.2 outlines five basic control parameters.

Each of these parameters is discussed below, to emphasise particular benefits or disadvantages for residual soils. Figure 6.6 shows a particular compaction curve obtained for a given compaction process. It shows the familiar optimum density characteristic, but also indicates the variation in undrained shear strength of the soil.

Table 6.1. Methods of compaction control.

Method	Comments	Standard
1. *In situ density*		
Sand replacement	Preferred, most soils, slow	Yes
Core cutter	Some soils, slow	Special
Nuclear meter	Uniform soil, calibration, rapid	Yes
Balloon densometer	Awkward, unreliable, slow	Yes
Special tests	Measure mass and volume	Special
Reference tests	Laboratory or field standard	Yes/special
2. *In situ moisture*		
Oven drying	Slow	Yes
Microwave drying	Calibration, rapid	Yes
Burner/ignition	Unreliable, dangerous, fast	Yes/special
Nuclear meter	Uniform soil, calibration, rapid	Yes
Reference tests	Laboratory or field standard	Yes/special
3. *In situ strength*		
Trafficability	Compaction equipment, other vehicles	No/special
In situ CBR	Calibration difficult, special equipment	Yes
Penetrometer	Versatile, fast, calibrate to other characteristics	Yes
Shear vane	Fast, fine grained cohesive soils only, calibrate	Yes
Resilience	(Deflectometer) special equipment	Yes/special
4. *Permeability*		
Infiltration	Simple ponding test	No/special
Constant/falling head	Special equipment, slow	Yes
Drill/auger hole	Special equipment, slow	Yes
5. *Laboratory*		
UU triaxial strength	Fast, select test conditions	Yes
CU triaxial strength	Slow, select test conditions	Yes
Optimum values	Relate strength to density data	Yes
Size corrections	Allow for stone and size effects	Yes

Table 6.2. Control parameters.

1	In situ density as compacted
2	In situ moisture content as compacted
3	In situ strenght as compacted
4	In situ permeability as compacted
5	Laboratory strength properties correlated to in situ measurements

The strength and stiffness of the compacted product are usually of greatest concern for fill performance. Figure 6.6 shows that a particular strength can be achieved for a range of densities and moisture conditions. The drier of optimum that the soil is, the more likely it is to have a large void content which may result in undesirable performance characteristics. It can be seen that specifying more than one characteristic is necessary to define an acceptable range of compaction conditions.

In situ density as compacted
In principle this is simple and direct. However, accurate field measurement of volume is time consuming and subject to procedural errors. Indirect measurements

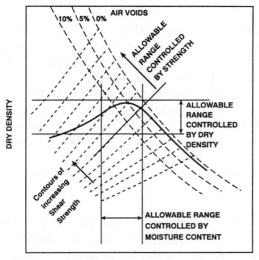

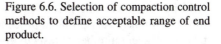

Figure 6.6. Selection of compaction control methods to define acceptable range of end product.

using nuclear moisture/density meters require careful calibration checks. This equipment may not be suitable for many remote sites.

The greatest disadvantage of density testing is that it offers only limited information on the fill properties that are really desired. The method of field compaction, and the particle sizes involved, may not correlate with the standard laboratory methods. At best, only inferences can be made about strength and permeability. Test results are compared to a laboratory value. Considerable material variability may occur in the field. Unless a corresponding laboratory reference test is performed for each field density measurement, there is a risk of field measurements being incorrectly interpreted.

In situ density has traditionally been the most popular method of measurement, due to its adoption and wide use for transported soils over a long period of time. For residual soils of high natural moisture content, experience relating to transported soils may be quite misleading. For residual soils, it is therefore recommended that other property measurements also be considered for control purposes.

In situ moisture content as compacted
Moisture content can be measured reasonably rapidly, and has the advantage that field variability can be assessed relatively easily by taking many measurements. Provided that the strength and permeability characteristics of the material are understood in relation to moisture content, moisture content can be used as an effective control parameter. However, the desired properties of strength and/or permeability are dependent on density as well as moisture content. The use of moisture content therefore has to be supplemented by measurement of other properties. Alternatively, if following a field trial, the method of compaction may be specified so as to guarantee achieving the requisite density, moisture content could be used as the sole control parameter. However, this would apply only in exceptional cases.

In situ strength as compacted

In principle, this is the most effective method of control, since strength is normally the most direct indicator of performance. Strength should be measured with a rapid test which is not subject to significant interpretation problems. A variety of rapid strength measurements can be used, ranging from hand-held or hand-operated vanes or penetrometers to in situ CBR tests and larger heavier penetrometers. Difficulties may occur if large particle sizes present in the soil compromise the performance and interpretation of the measurement technique. Technologically advanced compaction machinery or monitors, which use accelerometers to assess the response of the fill to vibration, must be calibrated to site conditions. Strength is also heavily dependent on moisture content. Thus moisture content must be used as a second control parameter.

In situ permeability as compacted

This is very effective where permeability is the most important characteristic for field performance. The greatest disadvantages are that field permeability testing is time consuming, very prone to procedural errors, and it may not be possible to under-take a sufficient number of tests to give a representative picture of field conditions.

Very crude and approximate field permeability tests, using simple infiltration methods, are easily performed provided that they can be done in areas which do not affect ongoing construction activities. In this case, permeability testing even if approximate, is a very useful guide.

Another disadvantage of permeability testing is that strength is usually also required. There are no effective correlations between strength and permeability, so that other forms of field tests are required anyway.

Laboratory strength properties correlated to in situ measurements

The advantages of laboratory testing relate to control and repeatability. Where this can be combined with an effective field measurement, very efficient compaction control can be achieved with a high degree of confidence.

The most widespread correlations are of laboratory strength related to field moisture content, for materials which have been adequately tested and whose variability is well understood. Thin-walled tube or core samples can be taken in the field, and tested at a site laboratory. The usual site laboratory strength test is the unconsolidated undrained (C_{uu} or 'quick') triaxial test.

SUMMARY

Because residual soils are derived from weathered rocks, variability of the soil will reflect variability of the parent rock. Variability of the source material may relate both to particle size distribution and to cohesive or frictional attributes. Compaction assessment should include monitoring of source materials to ensure that specification requirements can be met.

Field compaction trials are strongly recommended because they enable the work to be controlled as much as possible by factors that have been proven in the field. There is no problem with field trials being undertaken as part of the permanent works, but it is important to realise that trials can be slow and painstaking compared with full production. Both the supervisor and the constructor must recognise this and allow for it in their schedules and budgets.

The particular advantages of field trials are related to the practicalities of field conditions and to time constraints:

– Knowledge of compaction performance can be translated from laboratory test results, to results which directly reflect the performance of the compaction equipment. For example, field maximum dry density and optimum moisture content can be determined for the compaction equipment, and related to the number of roller passes, thickness of placed layers, etc.,

– Selection of lot sizes and testing frequency can be based on site conditions,

– Rapid field tests can be identified and selected so as to minimise the delays caused by waiting for laboratory test results,

– Method-based specifications can be proven and finalised. It is important that performance of method-based specifications be checked in the field. Methods of verification should be identified and recorded, and decisions made about the frequency of periodic laboratory or field tests which are still necessary as a check on the quality being achieved.

REFERENCES

Blight, G.E. 1962. Controlling earth-dam compaction under arid conditions. *ASCE Civ. Engin.* 54-55.

Blight, G.E. 1989. Design assessment of saprolites and laterites. Invited Lecture, Session 6. *12th Int. Conf. on Soil Mech. and Found. Eng., Rio de Janeiro* 4: 2477-2484.

Gidigasu, M.D. 1974. Degree of weathering in the identification of lateritic materials for engineering purposes – A review. *Engineering Geology* 8(3): 213-266.

Gidigasu, M.D. & Dogbey, J.L.K. 1980. The importance of strength criterion in selecting some residual gravels for pavement construction. *Proc. 7th Reg. Conf. Soil Mech. & Found. Eng. Accra, Ghana* 1: 317-300.

Rodda, K.V., Perry, C.W. & Roberto Lara, E. 1982. Coping with dam construction problems in a tropical environment. Engineering and Construction in Tropical and Residual Soils. *ASCE Geotech Div. Spec. Conf. Honolulu, Hawaii*: 695-713.

Terzaghi, K. 1958. Design and performance of Sasumua dam. *Proc. Inst. Civ. Engrs. London* 9: 369-394.

CHAPTER 7

Permeability

V.K. GARGA
Department of Civil Engineering, University of Ottawa, Ontario, Canada

G.E. BLIGHT
Civil Engineering Department, Witwatersrand University, Johannesburg, South Africa

7.1 PERMEABILITY CHARACTERISTICS OF RESIDUAL SOILS

Despite the enormous influence of seepage on slope stability, design of foundations for dams, excavations and underground openings, the geotechnical literature provides very limited information on the permeability of residual soils. The variation in the macrofabric of a weathering profile of a residual soil can result in large variations in permeability, both laterally and with depth. Generalizations of the 'typical' values of permeability for various types of residual soils can therefore be misleading and must be avoided. Lumb (1975) and Blight (1988) have indicated that the variation in the test results from a given site of residual soil in Hong Kong and South Africa can be of the same order of magnitude as the variation from site to site. Typical weathering profiles of residual soils presented by Lumb (1962), Deere & Patton (1971), Vargas (1974), Blight (1988) and others clearly indicate the variations in grain size, void ratio, mineralogy, degree of fissuring, and the characteristics of the fissures that will affect permeability values from site to site. Table 7.1 shows the relative permeability of weathering profiles in igneous and metamorphic rocks. Tables 7.2 and 7.3 show some values of permeability measured in undisturbed residual soils, both in situ and in the laboratory.

The methods used to determine the permeability of residual soils in practice both in the field and in the laboratory, are similar to those used for transported soils. The most common methods in the field are constant head and variable head permeability tests in boreholes, auger holes and test pits, (Brand & Phillipson 1985). Brand and Phillipson's review of international practice in testing of residual soils clearly indicates a strong preference for in-situ permeability testing. The limitations of small sized laboratory test samples to include the micro- and macro-structural variations encountered in the field are now clearly recognised.

The permeability of a saprolitic soil is controlled to a large extent by the relict structure of the material. Most of the flow takes place along relict joints, quartz veins, termite and other biochannels. De Mello et al. (1988) describe problems experienced with termite channels in the foundation of a 30 m high earth dam. Blight (1991) has reported on the effects of termite channels that were allowing effluent to

79

Table 7.1. Permeability of weathering profiles in igneous and metamorphic rocks (Deere & Patton 1971).

Zone	Relative permeability
Organic topsoils	Medium to high
Mature residual soil and/or colluvium	Low (generally medium or high in lateritic soils if pores or cavities present)
Young residual or saprolitic soil	Medium
Saprolite	High
Weathered rock	Medium to high
Sound rock	Low to medium

Table 7.2. Permeability values in residual soils of Brazilian Dam Foundations (Costa Filho & Vargas Jr. 1985).

Rock type	Residual soil	Permeability (m/y)	Type of test
Basalt	Mature residual and saprolitic	3×10^3 to 90	Variable head in piezometers and infiltration in boreholes
Basalt	Mature residual and saprolitic	30 to 0.03	Infiltration in pits and boreholes
Gneiss	Mature residual (porous clay) saprolitic	150 to 30	Infiltration and pumping in boreholes
Gneiss	Mature residual (porous clay) saprolitic	1500	Infiltration and pumping in boreholes
Gneiss	Mature residual (porous clay) saprolitic	70	Infiltration in pit
Gneiss	Saprolitic	30	Infiltration in pit and pumping
Gneiss	Saprolitic	30	Variable head laboratory permeameter
Migmatite	Saprolitic	1000 to 3	Infiltration in boreholes

Table 7.3. Permeability of residual soils derived from granitic and gneissic rocks.

	Rock type	Permeability (m/y)	Comments
Saprolitic soil	Granite	$125 \times 10^3 - 600$	Laboratory test
(young residual soil)	Granite	1000-0.15	–
	Granodiorite	0.15-175	Consolidation tests
	Granodiorite	3	Variable head permeability tests
	Quartz-diorite	3-100	Field and laboratory tests
	Gneiss	6-150	Laboratory tests parallel to schistosity
	Gneiss	3-60	Laboratory tests normal to schistosity
Mature residual	Granite	6-120	Laboratory tests
	Granite	60-0.15	–
	Granite	60-30	Micaceous layers in gneiss

leak from a series of evaporation ponds. In both cases the presence, or significance of these channels had been missed during the site investigation, and represented a very difficult repair problem.

Because the permeability is governed by macro-scale features, it usually cannot be reliably assessed by laboratory tests on undisturbed samples, as the scale of these is too small. The only completely reliable way is to assess the permeability by means of fairly large-scale field tests. If the water table is low, these can take the form of ponding tests or infiltration tests into test pits. Pumping tests from test pits or holes can be used if the water table is close to surface.

7.2 FIELD METHODS FOR MEASURING PERMEABILITY

The most common techniques to measure permeability in the field involve some form of either constant head or falling head testing in unlined or lined (cased) bore-holes. It is very common to obtain an estimate of permeability values by performing simple falling head tests in the drill stem at various depths as the drilling proceeds. In special cases, permeability testing can also be carried out using inflatable packers to isolate specific zones for testing, or by installing sealed hydraulic piezometers at different depths.

Many residual soils have sufficient cohesion to permit a testing hole to be opened up with a manual or machine operated auger without the need of a casing. A continuous-flight auger drill can be used successfully in unsaturated soils where speedy drilling may be necessary. Other methods for drilling in residual soils include wash borings, percussion-hammer drilling and rotary drilling. Whatever the method used, it is important that the inside surface of the hole used for permeability testing be free of loose or remoulded (smeared) material.

The most frequently used direct measurements of in-situ permeability can be divided into two major groups: those which feed water into the ground and those which extract water. The feed-in tests may be used above or below the water table, while the extraction tests can only be conducted below the water table. Because most residual soils occur in areas where water tables are relatively deep, extraction tests are seldom used to evaluate the permeability of residual soils.

O'Rourke et al. (1977) have compiled a very informative description of methods for in-situ measurement of permeability. The following provides a brief outline of the more common field methods used in practice. Reference to O'Rourke et al. is recommended for a detailed description of the test methods.

7.2.1 *Variable head tests*

Hvorslev (1951) remains the classic reference for the analysis of variable head tests in saturated soils, while Schmidt (1967) has provided a solution for falling head tests in unsaturated soil. However Schmidt's solution requires assumptions concerning the degree of saturation and the porosity of the soil. Hence the falling head tests can only provide a rough estimate of permeability for an unsaturated soil.

The method of analysis consists of first determining the basic time lag T, for which either of the two methods shown in Figure 7.1 may be employed. The coefficient of permeability k may be obtained from:

$$k \text{ (or } k_h \text{ if soil is anisotropic)} = A/FT \tag{7.1}$$

where A = cross-sectional area of the standpipe, F = shape factor shown in Figure 7.2 (F has dimension of length), k = isotropic permeability, k_h = horizontal permeability (in anistropic soil), T = basic time lag.

Where the soil is anistropic, the ratio of $k_h/k_v = m^2$ must be estimated, or obtained from laboratory tests. However, it should be noted that the error in evaluating the permeability of the soil due to error in selection of m is less than the inherent error in a falling head test.

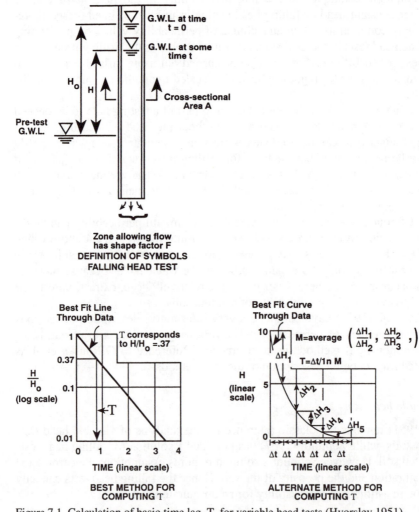

Figure 7.1. Calculation of basic time lag, T, for variable head tests (Hvorslev 1951).

i = isotropic conditions: $K_h = K_v = K$
a = anisotropic conditions: $K_h \neq K_v$

$$K \text{ or } K_h = \frac{A}{(F \cdot T)}$$

NOTE: Flow direction shown for Falling Head Tests for clarity;
"A/F" values also applicable for Rising Head Tests

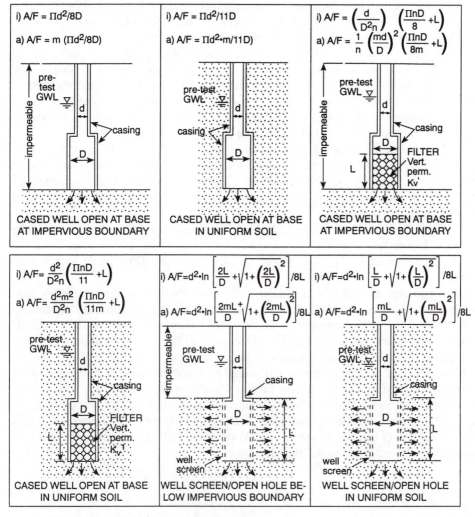

DEFINITIONS: $K_m = \sqrt{K_v K_h}$; $m = \sqrt{K_h / K_v}$; $n = K_v' / K_v$

where K_v = vertical permeability of soil/rock mass
K_h = horizontal permeability of soil/rock mass
K_v'= vertical permeability of filter in casing
T is termed the basic time lag.
See Figure 1. for best method to determine representative value of T

Figure 7.2. Shape factors for variable head tests (Hvorslev 1951).

7.2.2 *Constant head tests*

The analysis can be used for any feed-in test where the inflow during a test under a constant head becomes constant over time, i.e. when steady state flow conditions are achieved or approached. Methods of analysis include those by Hvorslev (1951), the US Bureau of Reclamation (Earth Manual (1974) and G-97 (1951)) and by Schmidt (1967). The Hvorslev analyses consist of the following steps:

1. Determination of the steady state conditions: An approximate value of the steady flow rate can be found from observed changes in flow rate with time, as follows: If H denotes the constant height of the water in the test hole above the base of the test zone, Q denotes the flow rate, and t denotes the time, then Q at steady state can be obtained from a plot of Q versus log $(1/\sqrt{t})$ or Q versus $1/\sqrt{t}$, as t becomes large. A typical plot of this kind is shown in Figure 7.3.

2. Determination of the effective head at test zone H_c: For all cases except for packer tests, this is the constant head of water above the test zone. In the case of packer tests, the height of the column of water above the test zone is adjusted for head losses in the water hose and couplings as well as for any additional pressure head supplied by the pump.

3. Determination of the shape factor F for a given test configuration: Use can be made of Figure 7.4.

4. Estimation of $k = Q_c/FH_c$ where Q_c is the constant flow under steady state conditions.

It is pertinent to note that the analysis of in-situ constant head permeability tests conducted in sealed hydraulic piezometers using Gibson's (1963, 1966) and Al Dhahir & Tan's (1966) approaches is also very common in practice. This analysis takes into account the compressibility of the soil and the resultant volume change as an excess (or deficit) pressure head is applied. A graph of the rate of flow Q versus $1/\sqrt{t}$ is plotted as the test progresses. It should be noted that it is not necessary to obtain a continuous record of the flow for the entire duration of the test. It is sufficient to monitor the flow rate over small time intervals periodically as the test proceeds. If t_1 and t_2 are the times (from commencement of the test) over which the flow rate is measured then Q can be plotted against $2/(\sqrt{t_1} + \sqrt{t_2})$. At large times, the expression for the coefficient of permeability reduces to:

$$k = \frac{Q_\infty}{F\Delta h} \qquad\qquad (7.2)$$

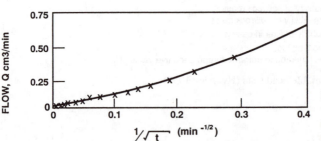

Figure 7.3. Typical plot of Q versus $\sqrt{t}$ from in situ constant-head permeability tests (Garga 1988).

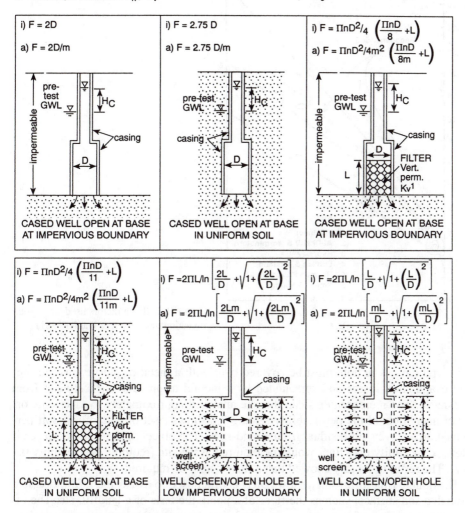

i = isotropic conditions: $K_h = K_v = K$
a = anisotropic conditions: $K_h \neq K_v$

$$K \text{ or } K_h = \frac{Q_c}{(F \cdot H_c)}$$

DEFINITIONS: $K_m = \sqrt{K_v\,K_h}$; $m = \sqrt{K_h/K_v}$; $n = K_v'/K_v$

where K_v = vertical permeability of soil/rock mass
K_h = horizontal permeability of soil/rock mass
K_v' = vertical permeability of filter in casing

Figure 7.4. Shape factors for constant head tests (Hvorslev 1951).

where Q_∞ = flow rate as t becomes large, Δh = constant head applied during the test, F = Shape factor.

The shape factors for cylindrical tips, length L and diameter D, are shown in Figure 7.5.

Often only a rough preliminary estimate of the soil permeability is required, in which case useful results can be obtained from a simple soak-away test in a test-pit. The test-pit is filled with water, and the fall in water level is recorded over a period

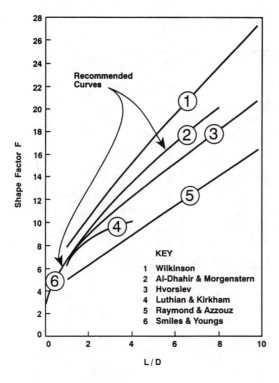

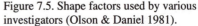

Figure 7.5. Shape factors used by various investigators (Olson & Daniel 1981).

of several days. The hole must be covered over with a metal or plastic sheet to prevent evaporation losses and surrounded by a mound to prevent surface water from running in. The measurements can then be analysed approximately to assess the order of magnitude of the permeability. Figure 7.6 shows a set of data of this sort performed on a clayey silt residual from a mud-rock. The object of the test was to see if a low-permeability clay liner would be required at the site (required permeability 0.1 m/y). The result of 24 m/y showed that a liner would be required.

The double ring infiltrometer test (DRI) is often used to assess the infiltration rate through a soil surface, such as the floor of a pond, the base of a landfill, etc. The infiltrometer consists of two concentric metal rings, open at the base, that are set into the soil. Usually, the edge of the metal ring is dug into the soil surface and the soil-metal interface is sealed by pouring a plaster of paris (gypsum) slurry around the edges of the rings, inside and out. Both rings are then filled with water to the same level and the rate of seepage from the inner ring is observed until it becomes constant. This may take up to three months to occur with highly impervious soil. The function of the outer ring is to remove edge effects from the inner ring and help ensure that seepage is vertically downwards. One of the problems of using this method is that the rate of evaporation may be of the same order as the rate of infiltration. Evaporation can be eliminated from the inside ring by covering it with a sealed plastic cover which is only briefly and partly removed to measure the water level, or to top up the water. To give representative results, the diameter of the rings should be as large as possible. Diameters of 1.2 m for the inner ring and 1.8 m for the outer ring are often used.

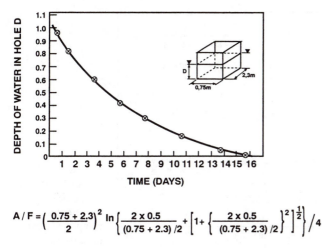

$$A/F = \left(\frac{0.75 + 2.3}{2}\right)^2 \ln\left\{\frac{2 \times 0.5}{(0.75 + 2.3)/2} + \left[1 + \left\{\frac{2 \times 0.5}{(0.75 + 2.3)/2}\right\}^2\right]^{\frac{1}{2}}\right\}/4$$

A/F = 0.36 m   T = 5.5 days

A/FT = 0.065 m/d = 7.5 x 10⁻⁵ cm/s = 23.7 m/y

Figure 7.6. Observed soak-away curve for soak-away test in a test pit. Approximate analysis according to case 5 in Figure 7.2.

More elaborate double ring infiltrometers (e.g. Daniel 1987) use a sealed bag to provide water to the inner ring. Figure 7.7 shows a set of infiltration results observed on a compacted clayey residual granite soil. It will be seen that although the rate of infiltration starts out at 1×10^{-5} cm/s, the final infiltration rate after 1700 h (70 days) is less than 1×10^{-7} cm/s. The steps in the experimental curve result because it is only practically possible to measure the water level to the nearest mm. The infiltrometer principle can be extended to ponding tests in which a pond measuring several metres square (e.g. 2 m × 3 m or 5 m × 5 m) is used to observe infiltration rates. In this case it is impractical to cover the surface of the pond to prevent evaporation. Evaporation losses can be estimated by means of a second pond lined with a geomembrane, or else the surface of the water in the pond can be covered with a film of lubrication oil to reduce evaporation. As with the double ring infiltrometer, edge effects can be eliminated by constructing a water-filled moat around the perimeter of the infiltration pond.

Figure 7.8 shows a set of observations made by means of three ponds, each measuring 3 m × 3 m. A fourth lined pond provided estimates of evaporation losses, and all four ponds were surrounded by a moat to eliminate edge effects. The soil was a silty sand residual from granite. There was a considerable scatter in the measurements with values on any particular day varying by factors of up to 6. The scatter appeared to result from the difficulty of measuring small changes of water level accurately and difficulty in compensating the measured seepages for evaporation losses which also depend on measuring small changes of water level. Temperature changes also affected the accuracy of the measurements by causing the plastic pipes, used as small stilling ponds against wind effects, to change in length. Thus an increase in temperature caused an apparent increase in seepage rate, and vice versa.

The determination of field permeability by Matsuo et al.'s (1953) method has been widely applied to residual soils, and especially to compacted residual soils in

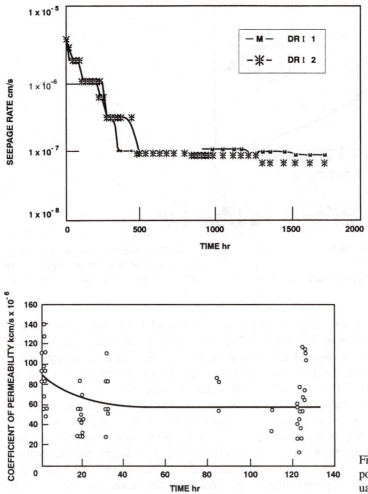

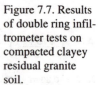

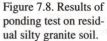

Figure 7.7. Results of double ring infiltrometer tests on compacted clayey residual granite soil.

Figure 7.8. Results of ponding test on residual silty granite soil.

earth dam construction. The test appears to work well in soils with permeability in the range of 10^{-4} to 10^{-6} cm/s (30 to 0.3 m/y). This simple method consists of excavating a large rectangular test pool, width B, length L. The sides of the excavation may be sloped if required. The flow rate Q necessary to maintain a constant level H in the pool is monitored. The seepage flow in this case is a three dimensional flow. In order to obtain a two dimensional estimate of the flow, the pool is next enlarged to twice the initial length ($L_1 = 2L$). The new flow rate, Q_1 required to maintain the same constant level H is noted. By subtracting the two flow rates, the effect of flow near both ends of the lateral cross-section can be eliminated, and the average discharge per unit length may be calculated as follows:

$$Q_{ave} = \frac{Q_1 - Q}{(L_1 - L)} \qquad (7.3)$$

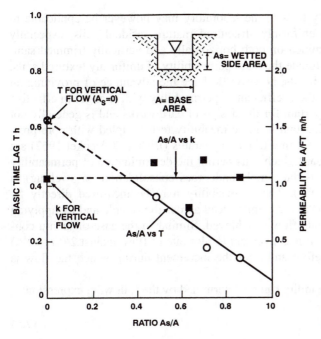

Figure 7.9. Variation of basic time lag T with ratio of side area As to base area A.

The range of in-situ permeability coefficient, depending on whether the flow is perpendicularly downwards or horizontal, may be obtained from the following simple expressions:

$$k = \frac{Q_{ave}}{B - 2H} \quad \text{and} \quad k = \frac{Q_{ave}}{B + 2H} \tag{7.4}$$

Figure 7.9 shows the results of a series of tests using Matsuo's method to explore possible anisotropy in permeability. As shown by the diagram, the calculated permeability proved to be constant regardless of the ratio of side area As to base area A of the pit. It could thus be concluded that the permeability was isotropic for all intents and purposes.

7.3 LABORATORY TESTS FOR PERMEABILITY

As mentioned earlier, permeability values determined in the laboratory do not necessarily represent the in-situ behaviour of residual soils. This is particularly true for undisturbed soils where the relatively small size of laboratory samples is inadequate to incorporate the various geological discontinuities, e.g. permeable and impermeable veins and fissures and other relict structures, present in a weathering profile. It should be noted that Darcy's law is only partially valid and that the coefficient of permeability of a soil can vary considerably with the flow gradient (Pregl 1987). It is therefore important to use a similar flow gradient in the laboratory to that likely to occur in the field.

Conventional permeability tests in the laboratory may however be applicable to compacted soils and more uniformly structured mature residual soils, especially where permeability is determined on both horizontally and vertically trimmed samples. It is then possible to estimate the mass permeability of uniformly textured soils. Laboratory tests, unlike field tests, however do have the advantage of providing an indication of the variation in the coefficient of permeability with changes in effective stress. This data is often important for the design of earthworks and is generally not available from field tests. Constant head permeability tests coupled with pore pressure dissipation tests carried out in a triaxial apparatus (Bishop & Henkel 1962) are particularly useful in this regard. Such tests permit the determination of permeability at various effective stresses as the sample undergoes consolidation (or swelling) in stages (Tan 1968, Garga 1988). The permeability may be measured directly by maintaining a small (10-20 kPa) differential head across the sample and by applying Darcy's Law when steady state flow is achieved. Similar to the case of in-situ constant head permeability tests, it is convenient to plot rate of flow against $2/(\sqrt{t_1} + \sqrt{t_2})$ where t_1 and t_2 are times before and after the increment during which the flow is measured.

From such a graph, permeability can be determined by the following expression:

$$k = \frac{Q_\infty L}{A\Delta h} \tag{7.5}$$

where Q_∞ = steady state rate of flow (volume per unit time), A = area of cross-section of the sample, L = length of sample, Δh = constant differential head across the sample.

Techniques to measure the permeability of unsaturated soil such as the instantaneous profile method, the pressure plate outflow test or the thermocouple psychrometer method are presently restricted to research use only. Reference may be made to Olson & Daniel (1981) and Hamilton et al. (1981) for an informative review of these methods.

7.4 COMPARISON OF PERMEABILITY MEASURED IN SITU AND IN THE LABORATORY

Although Darcy's law is usually applied to calculating flow through soils, there is considerable uncertainty as to how best to measure the Darcy coefficient of permeability k, as the following observations will show:

Day & Daniel (1985) conducted comparative field and laboratory measurement of permeability on two clays. Test ponds were constructed in the field, and samples were later retrieved from the test liners for laboratory measurements. Measurements of seepage rate were made for the pond as a whole, and by means of single and double ring infiltrometers. Tests using both rigid and flexible walled permeameters were made on block and tube samples of the clay compacted in situ, and also on samples compacted in the laboratory. Effective confining stresses in the laboratory were about 100 kPa and seepage gradients ranged from 20 to 200. Day and Daniel found that values of permeability deduced from seepage losses from the ponds were 900 to

2000 times larger than permeabilities measured in the laboratory, but only 1.2 to 1.9 times larger than field infiltrometer measurements.

Chen & Yamamoto (1987) also carried out a comparison of field and laboratory permeability measurements, using infiltrometers and porous probes in situ, and flexible-walled permeameters in the laboratory. For the laboratory tests, effective stresses were about 200 kPa and the seepage gradient was 180. They found field permeabilities were 10 times larger than laboratory values.

Elsbury et al. (1990) made a comparison of field and laboratory permeability measurements on a highly plastic clay. They found that double ring infiltrometer tests gave slightly lower permeabilities than did seepage rates from a test pond. However, compaction in the field with a vibratory roller resulted in a clay with a permeability ten times larger than one compacted using the same roller without vibration. Permeabilities measured in the laboratory used seepage gradients of 20 to 100 and effective stresses of 15 to 70 kPa. Permeabilities measured in the field proved to be between 10,000 and 100,000 times greater than values measured in the laboratory.

Pregl (1987) has stated that a permeability measured in the laboratory serves as an index of material quality but is not directly related to the permeability of a lining in the field. The permeability in the field will always be less than that measured in the laboratory, said Pregl, because the seepage gradient used in laboratory tests is usually of the order of 30 whereas that in the field approximates to unity. Also, the Darcy coefficient of permeability is not constant with seepage gradient.

It is apparent from these studies that there are several possible reasons why a permeability measured in the field may differ from one measured in the laboratory:

1. A large area exposed to seepage is more likely to contain defects in the form of more permeable zones than is a small area.

2. If the Darcy coefficient of permeability is not constant with flow gradient, the use of different seepage gradients in the field and laboratory will result in different field and laboratory values.

3. A similar remark applies to effective stresses. A specimen subjected to a high effective stress can be expected to show a lower permeability than a similar one with a low effective stress.

4. The interpretation of the field permeability test: The literature records various methods for calculating k. The most common is to assume that the seepage gradient $i = 1$, in which case $q = k$. Daniel (1987) uses the expression

$$k = \frac{qd_C}{(d_L + d_C)} \tag{7.6}$$

where d_C is the depth of soil being tested, and d_L is the depth of liquid in a ponding or double infiltrometer test.

To get some notion of the errors involved in this expression, suppose that $d_L = d_C/10$. If $i = 1$ or $d_L = 0$, $k = q$, and if $d_L = d_C$, $k = 0.5 \, q$.

Hence there is an appreciable measure of uncertainty in interpreting the results of ponding tests.

7.5 THE DARCY COEFFICIENT OF PERMEABILITY

According to the classical form of the Darcy Equation (7.5), one is led to believe that k is a constant for all i. Pregl (1987) has pointed out that this is not always so, but rather that k may increase with increasing i. The set of measurements shown in Figure 7.10 confirm this observation. The soil was a clayey sand residual from weathered granite and the measurements show that the seepage flow rate increases at a greater rate than the hydraulic gradient. For this set of data, the value of k at a hydraulic gradient of 1 was 1×10^{-4} cm/s, while at a hydraulic gradient of 20, k was 3×10^{-4} cm/s. Hence when measuring permeability in the laboratory, a flow gradient that is as close to that which will prevail in the field should always be used.

Nevertheless, it is possible to obtain reasonable agreement between field and laboratory permeability tests, as shown by Table 7.4 which compares the results of the ponding tests shown in Figure 7.8 with a number of laboratory test results.

All laboratory permeability tests were of the constant head flexible wall triaxial type. The average effective stress was kept at 3 kPa for all tests and the seepage gradient at unity. This stress was the lowest value that could be controlled reliably in the laboratory, but was still much more than the effective overburden stress in the pond tests of 0.5 m (only 4 to 5 kPa).

Table 7.4 compares the permeability values measured in the laboratory on specimens compacted to the same dry density as the upper 150 mm of soil in the ponds.

Hence this set of measurements shows that it is possible, with correct interpretation and correct testing, to estimate in situ permeabilities very closely from the results of laboratory tests.

However, it must be noted that the permeability of this granite soil is higher than the permeabilities that were considered in Section 7.4. It is noted from the literature that discrepancies between field and laboratory permeabilities generally appear to increase as the soil becomes less permeable, and presumably therefore, discontinuities and defects may play a greater role in modifying the permeability of large volumes of the soil.

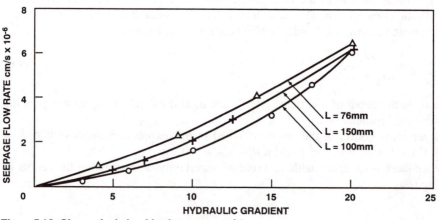

Figure 7.10. Observed relationships between q and i.

Table 7.4. Comparison of permeabilities measured in field and laboratory tests.

Range of Mean Permeability from Pond Test (cm/s × 10^{-6})	Range of Mean Permeability from Laboratory Tests (cm/s × 10^{-6})
59 to 81	37 to 93

It was also found in the series of tests described in Table 7.4 that double ring infiltrometer tests gave values of permeability that were considerably less than those recorded in the Table. The reason for this has not been discovered, but possibly also lies in a misinterpretation of the measurements.

REFERENCES

Al-Dhahir, Z.A.R. & Tan, S.B. 1968. A Note on one-dimensional constant head permeability test. *Géotechnique* 8: 4.

Bishop, A.W. & Henkel, D.J. 1962. *The Measurement of soil properties in the triaxial test*. London: Edward Arnold Publishers.

Blight, G.E. 1988. Construction in tropical soils. Keynote paper, *Proceedings, 2nd International Conference on Geomechanics in Tropical Soils, Singapore* II: 449-467.

Blight, G.E. 1991. Tropical processes causing rapid geological change. In *Quaternary Engineering Geology, Geological Society, Engineering Geology Special Publication* 7: 459-471.

Brand, E.W. & Phillipson, H.B. 1985. Sampling and testing of residual soils. A Review of International Practice. Southeast Asian Geotechnical Society, Scorpion Press, Hong Kong.

Chen, H.W. & Yamamoto, L.D. 1987. Permeability tests for hazardous waste management with clay liners. *Geotechnical and Geohydrological Aspects of Waste Management*, Lewis Publishers, USA, pp. 229-243.

Costa Filho, L.M. & Vargas Jr. E. 1985. Topic 2.3. Hydraulic properties. Peculiarities of geotechnical behaviour of tropical lateritic and saprolitic soils. Progress Report (1982-1985). Brazilian Society of Soil Mechanics, pp. 67-84.

Daniel, D.E. (1987). Earthen liners for land disposal facilities. *Proc.ASCE, Specialty Conference on Geotechnical Practice for Waste Disposal, ASCE, New York*, pp 21-39

Day, S.R. & Daniel, D.E. 1985. Hydraulic conductivity of two prototype clay liners. *ASCE Journal of Geotechnical Engineering* 111(8): 957-970.

Deere, D.V. & Patton, F.D. 1971. Slope stability in residual soils. *Proceedings 4th Pan American Conf. on Soil Mechanics and Foundation Eng. Puerto Rico* 1: 87-170.

de Mello, L.G.F.S, Franco, J.M.M & Alvise, C.R. 1988. Grouting of canaliculae in residual soils and behaviour of the foundations of Balbina dam. *2nd Int. Conf. on Geomechanics in Tropical Soils, Singapore* 1: 385-390.

Elsbury, B.R., Daniel, D.E, Sraders, G.A. & Anderson, D.C. 1990. Lessons learned from compacted earth liners. *ASCE Journal of Geotechnical Engineering* 116(11): 1641-1659.

Garga, V.K. 1988. Effect of sample size on consolidation of fissured clay. *Canadian Geotechnical Journal* 25(1): 76-84.

Gibson, R.E. 1963. An analysis of system flexibility and its effect on time-lag in pore water pressure measurements. *Géotechnique* 13(1): 1-11.

Gibson, R.E. 1966. A Note on the constant head test to measure soil permeability in-situ. *Géotechnique* 16(3): 256-257.

Hamilton, J.M., Daniel, D.E. & Olson, R.E. 1981. Measurement of hydraulic conductivity of partially saturated soils. *Permeability and Groundwater Contaminant Transport, ASTM STP746*: 182-196.

Hvorslev, M.J. 1951. *Time-lag and soil permeability in groundwater observations.* Bulletin No. 36, US Waterways Experiment Station, Vicksburg, Miss., USA

Lumb, P. 1962. The properties of decomposed granite. *Géotechnique* 12(2): 226-243.

Lumb, P. 1975. Slope failures in Hong Kong. *Quarterly Journal of Engng. Geology* 8: 31-65.

Matsuo, S., Hanmachi, S. & Akai, K. 1953. A field determination of permeability. *Proc. 3rd Int. Conf. on Soil Mech. and Found. Engng* 1: 268-271, Zurich.

Olson, R.E. & Daniel, D.E. 1981. Measurement of the hydraulic conductivity of fine-grained soils. *Permeability and Groundwater Contaminant Transport, ASTM STP746*: 18-64.

O'Rourke, J.E., Essex, R.J. & Ranson, B.K. 1977. Field permeability test methods with applications to solution mining. Report No. PB-272452, US Dept. of Commerce, National Technical Information Service.

Pregl, O. 1987. Natural lining materials. *Int. Symp. on Process, Technology and Environmental Impact of Sanitary Landfills. Cagliari, Italy* II(paper XXVI): 1-7

Schmidt, W.E. 1967. Field determination of permeability by the infiltration test. *Permeability and Capillarity of Soils. ASTM STP 417*: 142-158.

Tan, S.B. 1968. Consolidation of soft clays with special reference to sand drains. PhD thesis, University of London.

United States Bureau of Reclamation 1951. Permeability tests using drill holes and wells. Geology Report G-97 Denver, Colorado.

US Bureau of Reclamation 1974. *Earth Manual.* Water and Power Resource Service, US Dept. of the Interior, 2nd edition, Denver, Colorado, USA

Vargas, M. 1974. Engineering properties of residual soils from South-Central Region of Brazil. *2nd Int. Congress of the Int. Assoc. of Engng. Geology, Sao Paulo* 1: 5.1-5.26.

CHAPTER 8

Compressibility and settlement of residual soils

R.D. BARKSDALE
School of Civil Engineering, Georgia Institute of Technology, Atlanta, Ga., USA

G.E. BLIGHT
Civil Engineering Department, Witwatersrand University, Johannesburg, South Africa

8.1 COMPRESSIBILITY OF RESIDUAL SOILS

A number of methods have been used to assess the compressibility of residual soils. In situ methods have included the standard penetration test, the pressuremeter test and plate loading tests. Laboratory methods have been based on the oedometer and triaxial compression tests.

All residual soils behave as if overconsolidated. Their compressibility is relatively low at low stress levels. Once a threshold yield stress or equivalent preconsolidation stress has been exceeded, the compressibility increases. In most cases, the stress range will be such that the soil will remain within the pseudo-overconsolidated range of behaviour.

Figure 8.1 shows typical oedometer compression curves for a weathered andesite lava from the same profile as that represented by Figure 1.11. The yield stress is not very clearly defined, but correlates reasonably well with both depth and initial void ratio. Note the big variation of initial void ratio that may occur over a relatively small vertical distance. Figure 8.2 shows typical characteristics of the compression curve for a residual andesite lava in more detail. In particular, once the equivalent precon-solidation stress has been exceeded, a true preconsolidation stress can be established, and a residual soil will then behave similarly to a transported soil.

The equivalent preconsolidation pressure is probably a measure of the strength of the inter-particle, or intermineral bonds remaining in the soil after weathering. If this is the case, it would be reasonable to expect that σ'_{vc} would increase with depth in the profile. Figure 8.3 shows the variation of σ'_{vc} with depth for three profiles of andesite lava. These data show that σ'_{vc} does indeed tend to increase with depth and that the increase is roughly linear.

In a transported soil profile, lateral stresses are related to the overconsolidation ratio, and increase with increasing overconsolidation. As a residual soil weathers and decomposes, the minerals swell, but simultaneously lose material by leaching, internal erosion of ultrafine particles, etc. Hence unless the end-products of weathering are expansive, it is reasonable to expect that lateral stresses in a residual soil profile

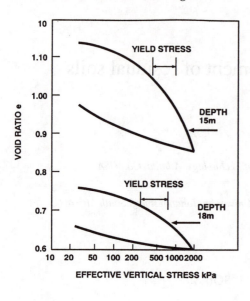

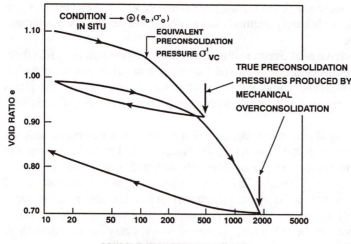

Figure 8.1. Typical oedometer curves for specimens from profile of residual andesite lava.

Figure 8.2. Typical consolidation curve for residual andesite lava showing the equivalent preconsolidation pressure and the establishment of true preconsolidation pressures by subsequent mechanical overconsolidation.

will be less than the overburden stress. In other words, K_0, the at rest pressure coefficient, will be less than unity.

An estimate of the lateral stress in a soil profile can be obtained by measuring the suction in an undisturbed specimen taken from the profile, and comparing this with the overburden stress. If σ'_s is the effective stress in the soil after sampling (which equals the suction in the specimen), it can be shown (Blight 1974) that

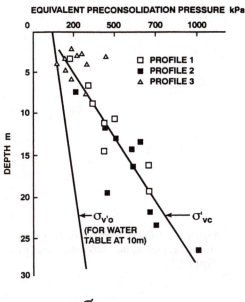

Figure 8.3. Variation of equivalent preconsolidation stress in three profiles of residual andesite lava.

$$R_s = \frac{\sigma'_s}{\sigma'_{vo}} = K_0 - A_s(K_0 - 1) \tag{8.1}$$

where σ'_{vo} is the effective overburden stress and A_s is the A-parameter for the change in pore pressure in the soil specimen that results from releasing the shear stress $\frac{1}{2}(\sigma'_{vo} - \sigma'_h)$ during sampling (σ'_h is the horizontal effective stress). As A_s is usually small and positive,

$$R_s = K_0 \text{ (approximately)} \tag{8.2}$$

Figure 8.4 shows that there is roughly a linear relationship between σ'_s and σ'_{vo}, for the profile referred to in Figure 8.3. Also that the value of R_s in the profile is about 0.3. The mean value for the well known Jaky expression for K_0 is

$$K_0 = 0.9 (1 - \sin\phi') = 0.38 \tag{8.3}$$

while the mean value of K_0 measured in triaxial tests in the laboratory is 0.42. Thus there is a strong indication that the value of K_0 for the profile is less than 0.5, which confirms the expectation deduced by considering the weathering process and products. While data of this sort appears to be available for only one profile (of andesite) it is not unreasonable to assume that K_0 in other profiles of non-expansive residual soil will similarly be less than unity, even though the soils behave as if overconsolidated.

Figure 8.5 shows the variation of void ratio e_0, compression index C_c and rebound index C_r in a profile of residual lava. There is a general tendency for all three parameters to decrease with depth (with a local anomaly at about 16 m) which indicates that the degree of weathering decreases with depth. The strong similarity in the shapes of the three curves suggests that both C_c and C_r should correlate with void ratio as indeed, Figure 8.6 shows they do. The data for a second andesite profile about 5 km away shows that the relationships in Figure 8.6 are of fairly general applicabil-

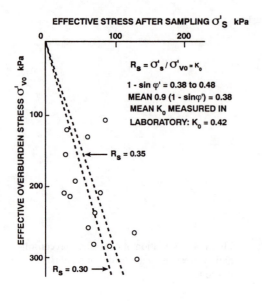

Figure 8.4. Indirect estimates of K_0 in a residual soil profile, the profile illustrated by Figure 1.11.

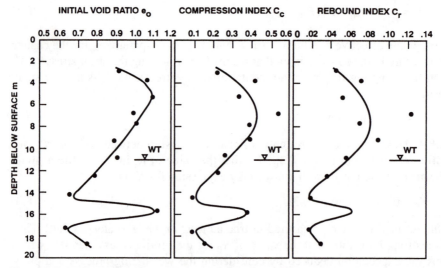

Figure 8.5. Variation of initial void ratio, compression index, rebound index and preconsolidation pressure with depth in profile of residual weathered andesite lava.

ity and that the void ratio of the weathered material gives a good indication of its compressibility.

Similar data are shown in Figure 8.7 for a norite gabbro that has weathered to a profile (illustrated in Fig. 6.1) consisting of a black highly expansive clay (or cotton soil) (usually from 1.5 m to 3 m in depth) overlying friable silty sands which grade with depth into very soft rock (usually between 3 and 9 m). These measurements by Hall et al. (1994) are very similar to those shown in Figure 8.6.

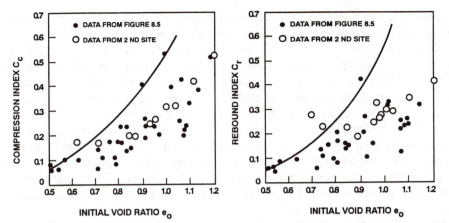

Figure 8.6. Correlations between initial void ratio and compression and rebound indices (Blight & Brummer 1980).

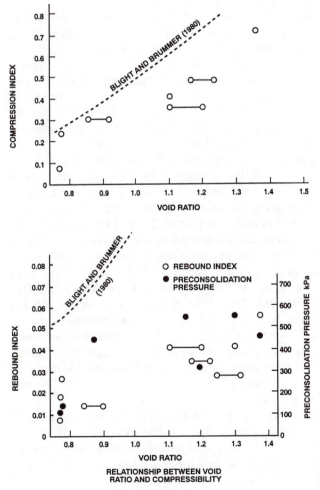

Figure 8.7. Hall et al's measurements on a saprolitic norite gabbro profile.

This brief introduction would be incomplete without mention of the activities and effects on soil compressibility of termites and other burrowing insects. Termites are very common in areas where residual soils occur, and their activities may significantly modify the compressibility of soils. There are two ways in which termite activity may affect the settlement of structures: In an area infested with termites either presently or in the past, the soil profile may be riddled with termite channels, thus materially increasing both the macro void ratio of the soil and its compressibility. Unfortunately, this effect does not appear to have been quantified, but it is an effect that should be recognised when examining, sampling and testing soil profiles. It is important to note that termite channels may be present in the soil even though there is no sign of termite mounds on the surface. The termites may have left the area decades or more ago, but the effects of their tunnelling will remain in the soil.

Because termites carry soil fines to the surface, large objects such as boulders and shallow foundations tend to be under-mined and carried downwards. In the case of a foundation, the settlement may be sufficient to cause structural distress. The following are a few examples of distress caused by termite activity:

– Partridge (1989) has reported that termite activity may result in the formation of a collapsible soil structure which may be subject to severe settlement if loaded during a dry period of the year and subsequently wetted by infiltrating water.

– In Johannesburg, one of a pair of augered cast in situ piles 10 m deep, supporting a pile cap, started to settle as load came on to it, with the result that the pile cap rotated. A 750 mm diameter auger hole was drilled next to the pile to determine the cause of the settlement. It was discovered that the pile had been founded directly above the food storage chamber of a termitary. This highly compressible spherical structure, about 500 mm in diameter was responsible for the settlement.

– In 1974 a precise survey base was constructed at Pienaarsriver, Transvaal by Finnish geodesists (Watt & Brink 1985). It was intended for the calibration of electro-optical and electronic distance measuring instruments to an accuracy of 1 part in 10 million. The monuments supporting the calibration marks were mass concrete blocks measuring 1 m square and founded at depths of 2.5 to 3 m on a yellow to reddish brown very stiff to very soft rock consistency silty sand, residual from the in situ decomposition of dolerite. Within two months it was found that certain of the monuments had settled by as much as 6 mm and had tilted slightly. This small movement could have been caused by shrinkage of the concrete, hence measurements were continued until 1976 when the precise distances were set out. At this time settlements of up to 16 mm had occurred. The movement continued and by 1980 it was found that the distance between the zero and 432 m marks had shortened by 12 mm (1 part of 36,000). The base was quite clearly unable to meet the accuracy requirements of 1 in 10^7 and had to be abandoned. Figure 8.8 shows settlement records for some of the monuments, indicating that settlements of more than 150 mm had taken place over a period of 11 years.

There was abundant surface evidence of termite activity in the area and test pits showed the existence of subterranean cavities, channels and food stores. There was evidence on surface that termites were actively transporting soil from below and depositing it on surface. Hence it is important to look for and, if found, record the presence of termites or termite channels in the soil.

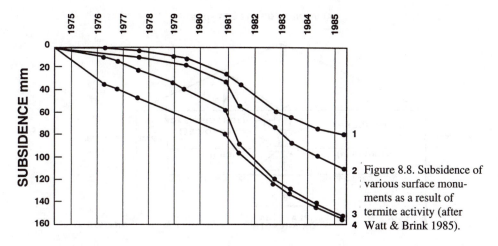

Figure 8.8. Subsidence of various surface monuments as a result of termite activity (after Watt & Brink 1985).

8.2 MEASURING THE COMPRESSIBILITY OF RESIDUAL SOILS

A thorough foundation investigation must be conducted in residual soils since they often vary erratically both with lateral position and depth. Test borings, advanced by either machine or hand, should be placed no further apart than about 20 m for commercial buildings and about 45 m for large industrial facilities. The groundwater table should be measured at the time the test boring is made and at least 24 hours later. These observations should be confirmed by means of carefully installed standpipe piezometers, observed over the ensuing months.

The following are the most commonly used in situ test methods to assess the compressibility of residual soils. As indicated previously, laboratory oedometer tests are also used very extensively. In selecting a test, the convenience, economy and good control over moisture and stress conditions of the oedometer test have to be considered in relation to the advantages (though at greater cost) of in situ tests. Further details of these tests are given in the appendix.

8.2.1 *Plate load test*

The plate load test is carried out by applying load to a rigid plate and measuring the resulting vertical deformation (Fig. 8.9). Since the plate load test is performed in situ, soil disturbance effects, which are important in residual soils, are reduced. Elastic moduli are calculated from the equation

$$E = \frac{qB(1 - v^2)}{\rho_0} I_f \tag{8.4a}$$

In which q is the stress applied to the plate, B is its width, ρ_0 is the corresponding settlement under the plate surface, v is Poisson's ratio (usually taken as 0.2 to 0.3, and values of I_f are given in Table 8.1. The elastic moduli obtained from the plate load test are usually greater than those determined from the laboratory consolidation tests even on specimens hand trimmed from block samples. Some studies have also

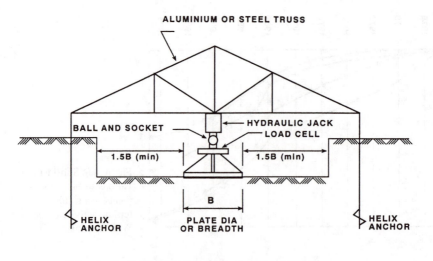

ALUMINIUM OR STEEL TRUSS

BALL AND SOCKET

HYDRAULIC JACK
LOAD CELL

1.5B (min)

1.5B (min)

B

HELIX
ANCHOR

PLATE DIA
OR BREADTH

HELIX
ANCHOR

Notes: 1. Conduct plate load test at several levels below the bottom
of the footing.
2. Test depth of influence is only 1 to at most 2 times the
plate diameter or width.
3. Plate deformations should be measured with at least 2 dial
indicators from an independant reference beam extending at
least 2.4m from the plate on each side.

Figure 8.9. Plate load test set-up.

Table 8.1. Influence factors I_f for computing immediate surface settlements ρ_o of shallow foundations.

Shape of the foundation					Rigid footing
		$$\rho_o = qB\left(\frac{1-v^2}{E}\right)I_f$$			
		Flexible Footing			
		Centre	Edge	Corner	
Circular		1.00	0.637	–	0.785
		(1.27)	(0.81)		
Rectangular footing	L/B	–	–	–	–
	1.0	1.12	0.76	0.56	0.815
		(1.37)	(0.93)	(0.69)	
	1.5	1.36	–	0.68	1.01
		(1.35)		(0.67)	
	2.0	1.52	1.12	0.76	1.12
		(1.36)	(1.00)	(0.68)	
	2.5	1.68	–	0.84	1.21
		(1.39)		(0.69)	
	3.0	1.78	–	0.89	1.30
		(1.37)		(1.68)	
	4.0	1.96	1.52	0.98	1.43
		(1.37)	(1.06)	(0.68)	
	5.0	2.10	1.68	1.05	1.55
		(1.35)	(1.08)	(0.68)	

Table 8.1. Continued.

10.0	2.56	2.10	1.28	2.10
	(1.28)	(1.05)	(0.64)	

Notes:
1. Assumptions – isotropic, homogenous, elastic semi-infinite soil mass,
2. B = diameter of circular footing and minimum dimension of rectangular footing; q = net increase in applied pressure,
3. Elastic Constants: E = modulus of elasticity, and v = Poisson's ratio
4. S (flex, rect) = $\lambda \cdot S$ (rigid, rect.), where λ= numbers given in parentheses,
5. For footings, edge of foundation is midpoint of long side.

found elastic moduli derived from plate load tests to be greater than those obtained from screw plate tests. (See the appendix for further details of the plate load test as well as for Table 8.1).

8.2.2 Cross-hole plate test

The cross-hole plate test is a variant of the conventional plate loading test. Two plates are used and are jacked horizontally against the sides of a test hole or trench. For reasons of greater safety, the test should only be performed after the installation of side support in the test hole. The load versus horizontal compression curve is measured, and the lateral compression is taken as half of the measured total extension of the jack.

The test is very convenient, as it can be used to measure elastic moduli at several depths in the same hole. The modulus measured is for horizontal compression of the soil, as is the case with the better-known pressuremeter test. The modulus for the horizontal compression does, however, appear to be not dissimilar to that for vertical compression at the same depth.

Figure 8.10 shows two stress-displacement curves recorded for cross-hole plate bearing tests in a weathered andesite lava.

Based on the assumption that each of the plates moves the same distance, a drained modulus of elasticity for the soil can be calculated using the expression of Bycroft (1956).

$$E'_h = \frac{(7 - 8v)(1 + v)\, \text{Pav}.\pi a}{16(1 - v)\rho_h} \tag{8.4b}$$

Where Pav = the pressure on the plate, ρ_h = the movement of the plate under load, a = the radius of the plate, v = the Poissons ratio (taken as 0.2 or 0.3), E'_h is the elastic modulus in the horizontal direction.

A set of typical results for a set of cross-hole jacking tests is given in Table 8.2.

8.2.3 Screw plate test

The screw plate test is a form of in situ plate load test that is performed at different depths beneath the surface. The screw plate consists of a single turn of a helical auger. Screw plate diameters typically vary from 100 to 300 mm. The screw plate

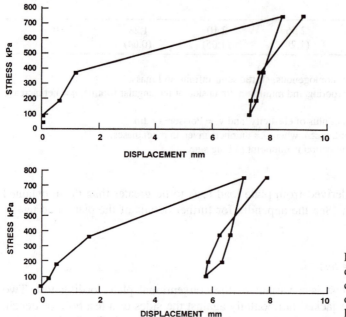

Figure 8.10. Stress-displacement curves for cross-hole jacking tests on a residual andesite lava soil.

Table 8.2. Set of typical cross-hole jacking test results.

Hole no	Depth (m)	Soil horizon	Field consistency	E'_h (MPa)	Stress range (kPa)
TP1	2.5	Res. Andesite	Firm	38.4	92 – 367
				8.1	367 – 736
TP2	2.96	Res. Andesite	Firm	30.5	46 – 367
				10.6	367 – 736
TP4	2.8	Res. Andesite	Firm	36.1	46 – 367
				27	367 – 736
TP7	2.4	Res. Andesite	Firm	40.5	46 – 367
				10.2	367 – 736

test is performed by screwing the plate into the soil, either by hand or machine, using a relatively simple jack and reaction system to apply load to the plate. When frictional resistance to turning the plate becomes too large, a hole the same diameter as the plate is augured out, and the plate is screwed below the bottom of the hole. The modulus of elasticity is then back calculated from the load-settlement curve. Valuable information concerning the rate of consolidation and the drained or undrained shear strength can also be estimated from screw plate test results. The screw plate test is described in detail, for example, by Selvadurai et al. (1980) and Smith (1987a, 1987b). The principle of the screw plate is shown in Figure 8.11 and further details of the test are given in the appendix.

Typical stress-settlement curves for a set of three screw plate tests are shown in Figure 8.12. These curves illustrate the difficulty often experienced in interpreting in situ tests on residual soils. As the depth of influence of a loaded foundation increases

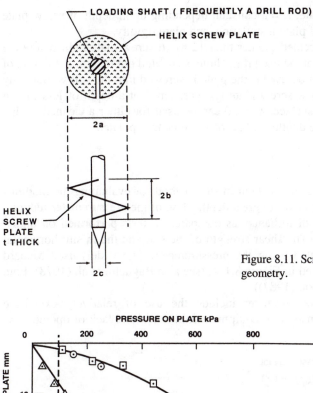

LOADING SHAFT (FREQUENTLY A DRILL ROD)

HELIX SCREW PLATE

2 a

2 b

HELIX
SCREW
PLATE
t THICK

2 c

Figure 8.11. Screw plate test – Helix screw plate geometry.

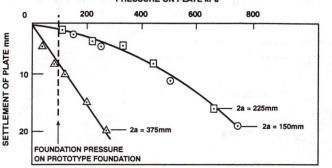

Figure 8.12. Results of screw-plate bearing tests on sand residual from weathered granite. All tests were at the same depth of 3m.

with increasing size, it would be expected that the settlements recorded on plates of increasing size would, at the same applied pressure, increase in proportion to their lateral dimensions. The set of results in Figure 8.12 does not show this trend, as a result of variability of the soil from the site of one plate test to another. Variability of this sort must be expected when testing residual soils, as it affects the reliability of the settlement predicted for the prototype structure.

Elastic modulus. The elastic modulus is calculated from the screw plate test results using the following general formula which is based upon the theory of linear elasticity:

$$E_{Sp} = \frac{\lambda \Delta q a}{\rho} \qquad (8.5)$$

where E_{Sp} = modulus of elasticity obtained from the screw plate test, Δq = net increase in average stress applied to the screw plate, ρ = measured screw plate deflection, a = radius of screw plate, λ = a constant depending upon depth of screw plate below the surface, method of plate installation, and plate rigidity.

For a deep screw plate located greater than 12 to 16 screw plate radii below the surface, a value of λ= 0.65 can be used if the hole is augered out to within 400 mm of the screw plate. λ = 0.75 can be used if the plate is screwed into the soil without any removal of soil. For a shallow screw plate test performed at a depth of less than 4 screw plate radii below the surface, λ = 1.0 can be used for either a cleaned or disturbed hole. For intermediate depths, values of λ can be interpolated.

8.2.4 *Pressuremeter test*

The pressuremeter test offers an excellent in situ method for evaluating the modulus of elasticity from the surface down to great depths. Use of the pressuremeter reduces, but does not eliminate, soil disturbance, as compared to tests performed on undisturbed, thin-walled samples. The shear strength of the soil and the in situ horizontal pressure can also be estimated using the pressuremeter. The widely-used Menard pressuremeter test is illustrated in Figure 8.13a. (see also Baguelin et al. (1978), Finn et al. (1984) and Mair & Wood (1987)).

Disadvantages of the pressuremeter include the use of relatively expensive equipment which is quite sensitive to equipment calibration and lack of operator ex-

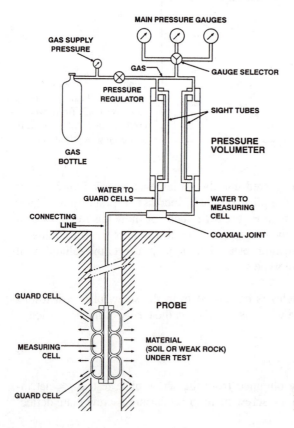

Figure 8.13a. Principle of Menard Pressuremeter test.

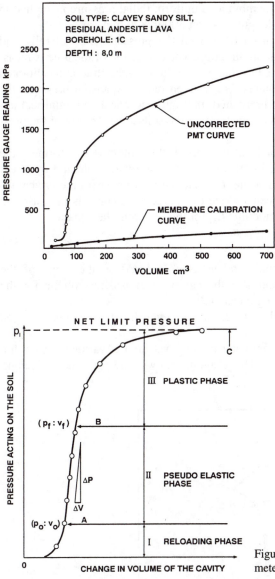

Figure 8.13b. Typical pressuremeter test results.

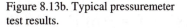

Figure 8.13c. Idealized Menard pressuremeter curve.

pertise. Pressuremeter test results are also influenced by the method of installing the device, test procedures, and the method of interpretation of the test results. Although the modulus of elasticity is measured in the horizontal direction, this does not appear to be an important disadvantage for residual soils.

The pressuremeter test can be visualized as a generalized form of lateral plate load test. The pressuremeter, which is cylindrical in shape, is inserted into a vertical borehole. The pressuremeter is expanded radially by increasing the fluid pressure inside the device. Cables and/or pipes are used to lower the pressuremeter down the hole and to connect it to instrumentation located at the surface. Since the outside of the pressuremeter consists of a rubber membrane, a uniform radial pressure is applied to the sides of the hole over a vertical length of about 300 to 800 mm.

Either gas or fluid pressure is used to expand the pressuremeter laterally and hence expand the borehole in the radial direction. The expansion of the hole is measured indirectly by either determining the change in volume of the fluid in a calibrated chamber or else by deformation gauges. For a given radial expansion pressure, the change in diameter of the hole can be related, through theoretical and empirical approaches, to the horizontal modulus of elasticity of the soil in the vicinity of the pressuremeter.

Because the Menard test is carried out in a pre-drilled hole, the soil forming the walls of the hole has been disturbed both by the forming of the hole and by stress release. To eliminate this effect, it is usual to cycle the pressure, with the typical result shown in Figure 8.13d. The steeper slope of the recycled curve is then used to calculate the pressuremeter elastic modulus E_M for the soil from the equation :

$$E_M (1 + v). \, 2V \, \Delta p / \Delta V \qquad\qquad (8.6)$$

where $v =$ Poisson's ratio, is usually taken as 1/3, $\Delta p / \Delta V =$ the slope of the (reloading) pressuremeter curve, and $V =$ the cavity or expansion volume for the pressure range over which E_M is being measured.

The recycled modulus is typically 2 to 2.5 times larger than the modulus for first loading.

Figure 8.14 shows a set of pressuremeter curves measured at various depths in a profile of weathered andesite lava. The diagram shows how the shear strength

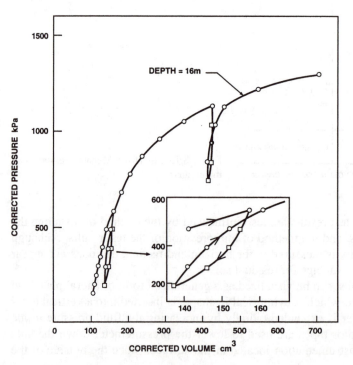

Figure 8.13d. Effect of cycling pressure to reduce effects of soil disturbance.

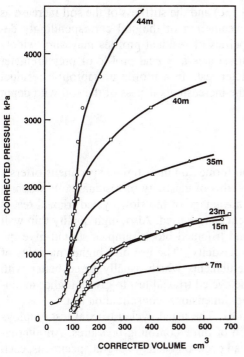

Figure 8.14. Pressuremeter curves measured in a profile of residual andesite lava.

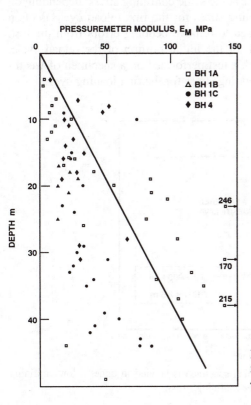

Figure 8.15. Variation of pressuremeter modulus of residual andesite lava with depth in 4 adjacent test holes.

(related to the net limit pressure in Fig. 8.13c) and the stiffness of the soil increase as the depth increases and the degree of weathering of the soil correspondingly decreases. As emphasized earlier, the properties of residual profiles may vary widely from one hole to another. Figure 8.15 illustrates a typical profile of pressuremeter moduli measured by Menard pressuremeter tests in a profile of saprolitic residual diabase. The profile shows the effect of the increasing stiffness of the soil with depth that was also illustrated by Figure 8.14.

8.2.5 *Slow cycled triaxial test*

A multi-stage, slow cycled triaxial test performed on undisturbed specimens offers a practical approach for evaluating the modulus of elasticity of residual soils when in situ tests are not feasible. An important advantage of the slow cycled triaxial test is that conventional triaxial testing equipment can be used. Also, high quality thin wall tube samples can be tested although hand trimmed block samples should give the best results (i.e. the highest modulus of elasticity). The use of a higher modulus of elasticity results in smaller estimated settlements which usually agree better with measured values. One study has found the cycled triaxial test to give a higher modulus of elasticity than the conventional, one-dimensional consolidation test.

Further details are given in the appendix. The usual cycled triaxial test employs two complete load-unload cycles for each of three confining pressures. An alternative method is to do one load-unload-reload cycle on each of several specimens, each at a different confining stress, or a number at the same confining stress, depending on the availability of specimens. The confining stress for the first reload cycle is often the nominal 'bedding' value which is used to remove bedding errors from the test. The elastic modulus is then determined for the initial portion of the reload cycle. Figure 8.16 shows a typical result of such a test performed on a specimen of weathered andesite lava, and using an initial bedding stress for the first loading cycle.

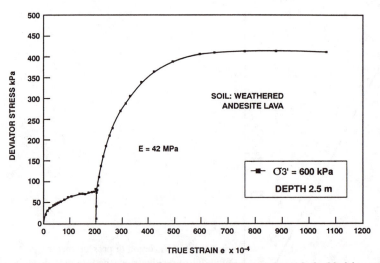

Figure 8.16. 'Bedding cycle' triaxal test in which specimen is bedded in under a low confining stress, unloaded and then reloaded under a higher confining stress.

8.2.6 *Comparisons of different methods of assessing elastic modulus for residual soils*

Very few comparisons have been published of the various available methods of assessing compressibility or elastic moduli for residual soils. One of the few such comparisons was made by Jones & Rust (1989) for a saprolitic weathered diabase. Their comparison, illustrated in Figure 8.17, shows that the self-boring pressuremeter, plate bearing test and oedometer test give comparable results for E provided that rebound curves are used for the plate loading and oedometer tests. Also note from Figure 8.17 that the N value from the Standard Penetration Test (SPT) in the same profile has a very similar trend with depth. From this it may be deduced that SPT results in residual soils may also be used to obtain estimates of E. For this profile, the correlation is:

$$E = 1.6N \text{ MPa} \tag{8.7}$$

A popularly used correlation is

$$E = N \text{ MPa} \tag{8.8}$$

which is rather more conservative than Equation (8.7).

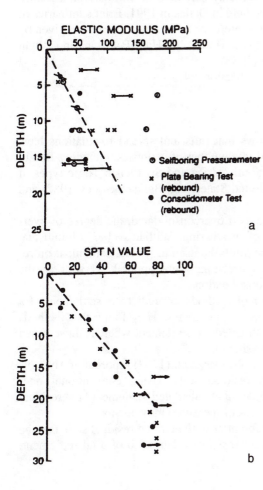

Figure 8.17. a) Comparison of E values derived by three different methods for a weathered diabase profile, b) Variation of standard penetration test N with depth in the same profile.

8.3 EXAMPLES OF OBSERVED SETTLEMENTS OF STRUCTURES FOUNDED ON RESIDUAL SOILS

8.3.1 *Shallow foundations*

The main problems with shallow, lightly loaded foundations on residual soils arise from seasonal or cumulative swelling or shrinkage or from collapse of highly leached, unstable grain structures on wetting. By their very nature, these problems tend to occur in the less humid tropical and sub-tropical areas where seasonal or perennial soil moisture deficits occur, usually in combination with active clays, highly weathered granites or loess-type soils.

The subject is a highly specialized one, and is of great importance, as expansive and collapsing soils necessitate many millions of dollars worth of repairs to homes throughout the tropics and sub-tropics annually. Very often, the most extensive damage is caused to low-cost homes whose owners can least afford the cost of repairs. Solutions to these problems tend to be highly site-and country-specific. The reader is referred to the voluminous literature on the subject, and in particular to the series of six international conferences on expansive clay soil held at various venues since 1965, culminating with the 7th conference held in Dallas in 1991. Brief summaries of the approach used in various countries and reference to literature have been given by Richards (Australia), Gidigasu (Ghana), Desai (India), Ul Haq (Pakistan) and Blight (South Africa), all in Brand & Phillipson (1985).

The subject receives further treatment in Section 8.4.

8.3.2 *Raft and spread foundations*

Brand & Phillipson's (1985) survey shows that rafts and spread foundations have been used to transfer load to residual soil in a number of countries, including Brazil, Hong Kong, India, Nigeria, Singapore, South Africa and Sri Lanka. These types of foundation are also widely used in the United States (e.g. Barksdale et al. 1982) and Australia (Moore & Chandler 1980).

As stated earlier, residual soils behave as if overconsolidated, the degree of overconsolidation depending on the degree of weathering. Williams (1975) found that conventional oedometer tests could successfully be used to predict settlement on residual soils provided the Schmertman corrections were applied, as well as the Skempton-Bjerrum corrections for overconsolidation.

Barksdale et al. (1982) used a number of methods to predict the settlement of a large water tower founded on a weathered biotite gneiss. They found that methods based on in situ tests considerably overestimated the settlement whereas those based on laboratory tests could give excellent predictions.

Working in the same geographical area, Wilmer et al. (1982) reported on five settlement studies of spread footings on residual soils. Using conventional one-dimensional oedometer tests, they concluded that calculated settlements for residual soils would normally be about 30% in excess of measured movements.

Barksdale et al. have observed that differential settlement on residual soils may be as much as 75% of total settlement, but averages less than 25% of total settlement.

Their observations tend to be rather more favourable than those of Burland et al. (1977) for differential settlements of buildings on transported clays.

Two detailed settlement studies have been made for structures founded on weathered andesite lava in Johannesburg, South Africa, (see Fig. 8.18): On the residual profile described in Figure 1.11. Jaros (1978) used a pseudo-elastic finite element method to predict the settlement of the two multi-storey structures with reasonable success. Recognizing that the material behaved as if over consolidated, he used rebound curves from oedometer tests to derive his pseudo-elastic constants. He did not however, attempt a full time-settlement analysis. The time settlement records for the two buildings, Total House and Guardian Liberty Centre which are both founded on rafts at depths of 18 m and 15 m respectively, are given in Figures 8.19 and 8.20. Both of these records show that most of the settlement occurred during construction with post-construction settlement amounting to only 10% to 20% of total settlement. If the settlement measurements had been commenced after construction, very little movement would have been recorded. As far as the amount of settlement is concerned both analyses proved reasonably accurate. In the case of Total House, Jaros overpredicted the actual settlement by 26%, while for Guardian Liberty centre, his calculation underpredicted by 21%. Jaros ascribed the discrepancy in the case of Total House to the presence of large quartzite inclusions or floaters (see Fig. 8.18) in the lava, the effect of which could not adequately be considered in the analysis.

Pavlakis (1983) re-analysed the settlement records for the two buildings. Using a conventional analysis based on pressuremeter tests, and correcting for the presence of the quartzite floaters, he was able to predict the measured settlements very closely, as recorded on Figures 8.19 and 8.20. Intrinsically, however, Pavlakis' pressuremeter

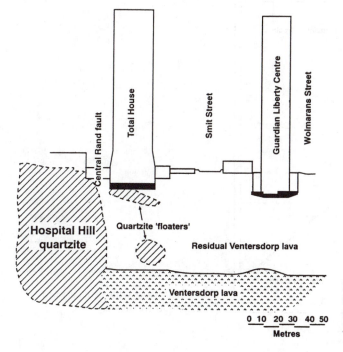

Figure 8.18. Section through the Total House and Guardian Liberty Centre buildings, Johannesburg (Jaros 1978).

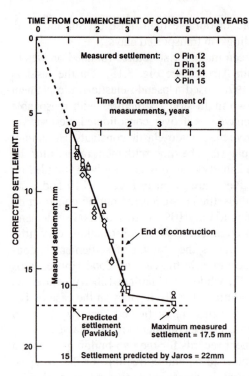

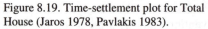

Figure 8.19. Time-settlement plot for Total House (Jaros 1978, Pavlakis 1983).

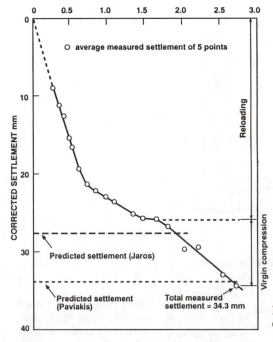

Figure 8.20. Time-settlement curve for Guardian Liberty Centre.

prediction was no more accurate than Jaros' prediction based on oedometer tests. The difference in the predictions arises from Pavlakis' adjustment of the time zero for the two time-settlement curves. Nevertheless, these two independent analyses illustrate that the settlement of structures on residual soils can be predicted with adequate accuracy for most engineering purposes.

8.3.3 *Deep foundations*

According to the Brand and Phillipson survey, deep foundations of various types are widely used in residual soils. Driven displacement piles and driven steel tube piles have been used in Brazil, but bored piles and caissons of various types appear to be more widely used in tropical soils. Hand-dug caissons are widely used in Hong Kong, with bored contiguous piles frequently used to support the sides of building excavations. Bored piles are also used in India, Nigeria and Singapore. Driven *H* piles and precast concrete piles are used in Singapore. In South Africa the commonest type of pile in residual soils is the bored cast in situ pile, although driven and driven displacement piles are also used. The situation in Sri Lanka is similar.

Pavlakis (1983) has had considerable success in predicting settlements of piles from the results of pressuremeter tests in weathered andesite lava. To predict the pile failure loads, he used standard Menard techniques (Menard 1975), while to predict the load-settlement curve, he followed the procedure of Sellgren (1981). Sellgren's method for calculating settlements is briefly described in the appendix. Figures 8.21a and 8.21b are examples of the excellent correlations he obtained between measured and predicted load-settlement curves, in Figure 8.21a for a single driven cast-in-situ pile, and in Figure 8.21b for two piles loaded through a single pile cap.

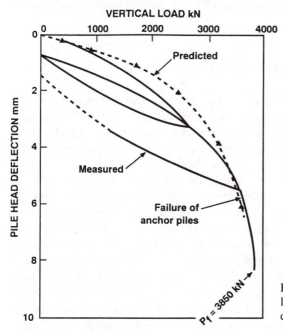

Figure 8.21. a) Measured and predicted load-deflection behaviour for a single driven cast-in-situ pile.

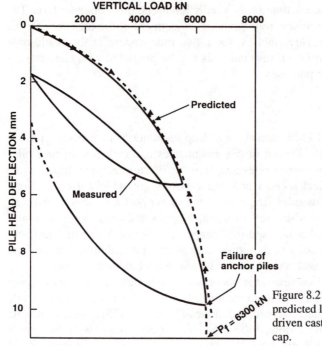

Figure 8.21. Continued. b) Measured and predicted load-settlement behaviour of 2 driven cast-in-situ piles under a one pile cap.

8.3.4 *Conclusions*

The comparisons of predicted and measured settlements given above show that the methods adapted for use with residual soils are effective and sufficiently accurate for use in most design.

8.4 HEAVE AND COLLAPSE SETTLEMENT

The main objective of this chapter has been to describe the volume change and settlement characteristics of residual soils. However, the chapter would not be complete without mention of two phenomena that very commonly occur with residual soils in semi-arid climatic zones – the problems of heave (or swell) and collapse settlement.

In terms of the unsaturated effective stress variables $(\sigma - u_a)$ and $(u_a - u_w)$ (e.g. Blight 1965) the processes of heave and collapse are illustrated by Figures 8.22a and 8.22b. Figure 8.22a shows the relationship between $(\sigma - u_a)$ and $(u_a - u_w)$ for the swelling process in an unsaturated soil. If $(\sigma - u_a)$ is kept constant, the equivalent of a constant total stress in the field, a reduction of the suction $(u_a - u_w)$ resulting from increasing moisture will cause the soil to swell along a line such as AB in Figure 8.22a. If the soil has a stable fabric or grain structure, swell will continue until $(u_a - u_w) = 0$ and the soil becomes saturated (at B). Thereafter, swell will continue along BC', if σ is further reduced (u_a does not now exist, as the soil is saturated). If, however, the soil has an unstable grain structure, collapse settlement may occur once $(u_a - u_w)$ falls below a critical value for the value of $(\sigma - u_a)$ being carried, (path DEF).

Once the grain structure has stabilized, swell will resume along FG, if $(u_a - u_w)$ continues to be decreased, and the soil will behave normally.

Figure 8.22b illustrates the results of a constant volume swell process in which $(u_a - u_w)$ is reduced by increasing the moisture content of the soil, and $(\sigma - u_a)$ is adjusted to prevent volumetric strain or swell from occurring. The paths traced out in the plane of zero volumetric strain represent contours of constant effective stress and can be followed regardless of whether the soil is expansive or collapsing. At constant effective stress, collapse will not occur, regardless of the value of $(u_a - u_w)$.

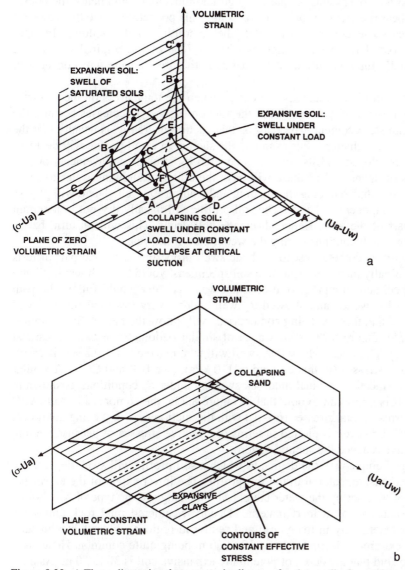

Figure 8.22. a) Three-dimensional stress-strain diagram for the swell of a partly saturated soil under constant isotropic load, b) Three-dimensional stress-strain diagram showing contours of constant effective stress in partly saturated soils.

8.4.1 *Heave or swell of residual soils*

Heave is usually experienced with soils residual from weathered shales and mudrocks as well as from basic igneous rocks such as dolerites, norite gabbros and some andesite lavas. In other words, heave is experienced with clayey residual soils, especially if the clay is smectitic. Damage to structures by the heave or expansion of clays occurs in many parts of the world, and is characteristic of arid and semi-arid zones. The structures most commonly damaged by heave (or sometimes shrinkage) are houses and other types of low-rise dwelling unit. Insurance policies usually specifically do not cover damage caused by ground movement, and hence the house-owner has to bear the cost of repairs directly out of his pocket. Damage by heave can be severe, even so severe as to necessitate the demolition of the building. In other cases, repair costs have been as high as 60% to 70% of the original cost of the structure. Usually there is no complete guarantee that the damage will not recur in the future.

As illustrated in Figure 8.22a, heave occurs when the total stress on the soil ($\sigma - u_a$) remains approximately constant, but the suction ($u_a - u_w$) decreases because the soil gains in moisture content. The reasons that suction decreases is usually that the surface land-use has changed. For example, the suction in a soil profile will decrease if the land is developed and its surface is covered by structures, pavements or irrigated gardens or lawns. All of these will have the effect of allowing water to accumulate in the soil, thus reducing the suction, and leading to swell. There will always be a seasonal component to heave. The soil profile will usually dry out to a certain extent over part of the year. Hence shrinkage will occur. This will, in turn, be reversed and heave will continue during the wetter part of the year.

In a laboratory oedometer test to evaluate swell properties, a typical pair of tests results for nominally identical expansive soil specimens would be as shown in Figure 8.23a. One specimen, compressed at constant water content would follow the path ABC. If the soil is wetted and allowed to swell under a very low total stress, it will swell along AA' and then, on being compressed, will follow the path A'BC (Jennings & Knight 1975). Figure 8.23b shows a set of similar oedometer tests on a residual weathered shale. These tests show that swell will take place even if the soil is carrying considerable stresses (in this case up to 110 kPa), (see BB' and CC'). (A similar test result for a smectitic residual mudrock swelling under K_o conditions, is shown in Figure 9.31). It is important to note that in Figure 8.23a, the compression path ABC is plotted in terms of total stresses ($\sigma - u_a$) with the suction ($u_a - u_w$) being unknown. The path A'BC, however, is plotted in terms of effective stresses σ' because wetting the soil will have reduced the suction to zero.

Other things being equal, the degree to which a soil profile will swell depends on the depth of expansive material in the soil profile. The lower extent of the expansive part of the profile is set by the water table, or by a change in the type of soil from a potentially expansive to an inert material. Deeply weathered mud rock and shale profiles occur extensively in many arid and semi-arid regions. Water tables in such areas may be extremely deep, depths of 30 to 50 m being quite common. Hence it is not unusual to find that the depth of potentially expansive soil is 30 to 50 m. As a result, amounts of heave can also be very large (hundreds of mm).

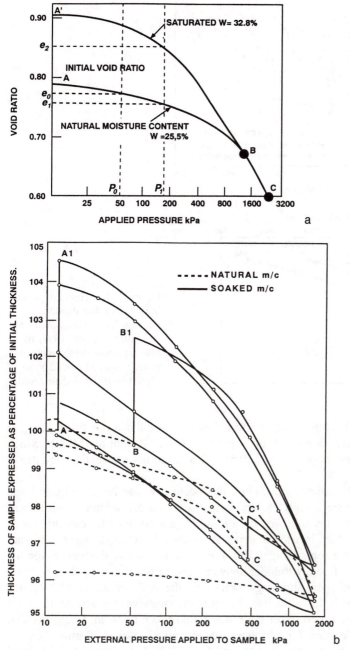

Figure 8.23. a) Compression curves for a heaving soil,b) Double oedometer test on residual weathered shale.

Williams (1991) for example, has recorded surface heaves approaching 500 mm. At the same site, he has observed heaves of 200 mm at a depth of 10 m below surface (see Fig. 8.24a). At this site, the depth of expansive material approaches 50 m. Usually, however, surface heaves are more moderate and seldom exceed 150 mm. Figure 8.24b shows a series of heave-depth curves for a group of houses supported on slabs-

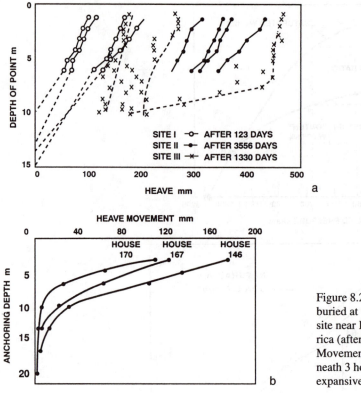

Figure 8.24. a) Heave of plates buried at various depths at a site near Kimberley, South Africa (after Williams 1991), b) Movements of depth points beneath 3 houses founded on an expansive residual shale profile.

on-grade on a residual weathered shale foundation where the water table was at a depth of 25 m. The heave at a depth of 3 m varied from 110 mm to 180 mm which translates to surface heaves of 180 to 220 mm. These are not unusually large heave movements.

The depth to which seasonal movement may occur can also be large in arid and semi-arid zones. There are usually clearly defined wet and dry seasons, with a short wet season of 4 to 5 months followed by a long dry season. Profiles can dry out to depths of 15 to 20 m, which is also the depth to which soil can be desiccated by suction originating from tree roots (Blight 1987). Rain at the start of the wet season often occurs in the form of heavy thunder showers, and infiltration may deeply penetrate the profile down open shrinkage cracks in the soil. Pellissier (1991), for example, has found free water at the base of a pile in expansive clay at a depth of 7.5 m, shortly after rain, where no water table was found down to 16 m. It appears that the heave of this profile from 7.5 m downwards amounted to over 70 mm, showing that the soil must have been desiccated to well below 7.5. The pile-head had heaved extensively and the pile was suspected of having failed in tension. The pile was progressively exhumed and its load transferred to three jacks at the ground surface. Figure 8.25 shows the load on the jacks and the inferred load distribution down the length of the pile. The interesting thing about the load distribution is that it appears that there was virtually no friction on the shaft of the pile down to a depth of 4 m.

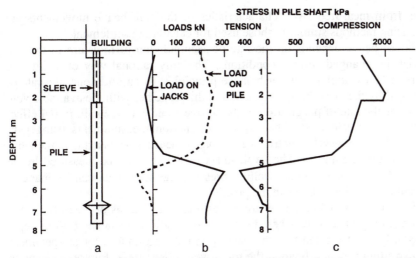

Figure 8.25. Measurements made on a pile in deep residual weathered mudrock. a) Profile of pile, b) Load on pile and jacks, c) Stresses in pile shaft (after Pellisier 1991).

The portion between 4 m and 5.5 m was subject to frictional uplift, and the length from 5.5 m to 6.7 m was acting to anchor the uplift. The upper 2.3 m of pile had been sleeved to reduce the friction, but the following 1.7 m must effectively have been out of contact with the surrounding soil as a result of seasonal drying. A similar observation has been made by Zeevaert (1980) relating to piles in Mexico City.

In another case, where the water table was relatively shallow (8.5 m), Blight (1965) observed an almost immediate seasonal heave response to rain on a depth point anchored at 5.5 m indicating that seasonal desiccation occured to below this depth.

Engineering solutions to counter the effects of heave are difficult to formulate. For heavy engineering structures, anchor piles or sleeved anchor piles (Blight 1984) may be appropriate. For light structures, Williams (1980) and Blight et al. (1991) have reported considerable success in using surface sprinkling or ponding to pre-heave a site and reduce differential movements. Prewetting to pre-heave the profile is therefore another candidate solution. Stiffened rafts appear to be yet another popular solution for light structures (e.g. Pidgeon 1980).

8.4.2 *The prediction of heave in residual soils*

In a comprehensive survey, Schreiner (1987) listed 39 published methods of predicting heave in expansive clay profiles. The procedures listed include completely empirical methods, usually based on indicater test results, methods based on void volume available to hold increased moisture, laboratory simulations of field processes, and a few applications of effective stress principles. The most fundamental method combines effective stress principles, void volume considerations and water balance principles.

The basis of method is as follows:

– The initial and final effective stresses and hence the changes in effective stresses in the soil profile are estimated.

– Hence, from measured swell characteristics of the soil, heave movements are calculated, using methods similar to those used for calculating settlement.

– The rate of heave depends on the rate of accumulation of moisture in the soil profile under the changed surface conditions. The only rational way of estimating this appears to be by applying water balance principles. In an arid zone, the rate of heave is limited by the availability of water, and this, together with the available void volume governs the rate of penetration of the 'heave front' into the soil. Initial effective stresses can be estimated either by in situ suction measurements using psychrometric methods, or psychrometric or filter paper measurements made in the laboratory on undisturbed samples. The swell index of the soil can be measured in the laboratory, and at the same time, swelling pressure measurements can be made to determine initial effective stresses in the profile.

Figure 8.26 shows initial effective stresses estimated by means of measured swelling pressures in a profile of weathered siltstone having a water table at 30 m depth (Blight 1984). It will be noted that the data is erratic, and quite a lot of specimens showed no swelling pressure beyond the total overburden stress. Figure 8.27 shows measured values of the swell index C_s and its variation with depth at the same site. Clearly, the expansiveness of the soil decreases with increasing depth, as does the potential for expansion, in terms of moisture stress.

Estimating the final effective stresses in the profile is even more difficult than estimating initial effective stresses, as these are dependent on the long-term soil microclimate. If the surface will be sealed, and it is unlikely that much water will be contributed to the soil via leaking sewers, soak-aways, etc., the approach first suggested by Russam & Coleman (1961), and developed by Aitchison & Richards (1965) can be used. The Russam-Coleman/Aitchison-Richards diagram (Fig. 8.28) relates the Thornthwate moisture index T to the equilibrium suction at depths of 450 mm and 3 m below the centre of a paved area. To obtain the complete profile of final effective

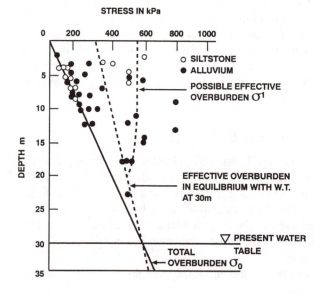

Figure 8.26. Effective stresses in a deep profile of weathered siltstone (containing alluvium-filled channels) estimated from swelling pressure tests (Lethabo power station).

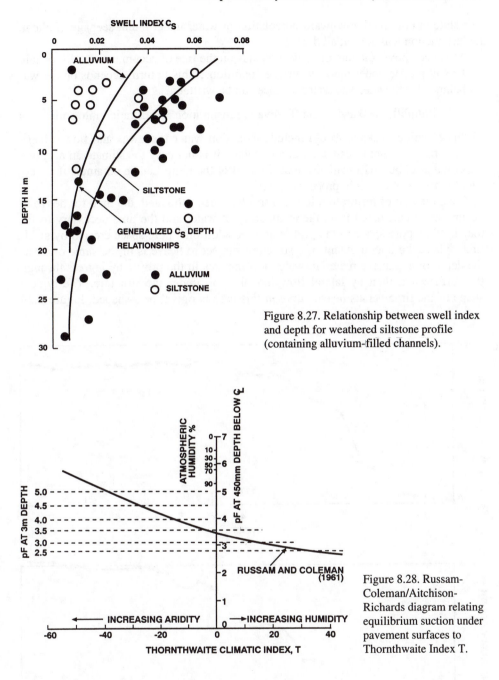

Figure 8.27. Relationship between swell index and depth for weathered siltstone profile (containing alluvium-filled channels).

Figure 8.28. Russam-Coleman/Aitchison-Richards diagram relating equilibrium suction under pavement surfaces to Thornthwaite Index T.

stresses, it is necessary to know, or assume the suction at some other depth, possibly from the position of the water table.

In the case from which Figure 8.26 was derived, the site of power station, observations after the power station was in operation showed that there were so many leaks and spillages of water at the surface, that the final suction profile corresponded to a state of continual downward percolation of water under a unit seepage gradient, i.e. the suction was zero at all depths.

As stated above, the rate of heave depends on the rate of accumulation of moisture in the soil profile under the new surface conditions. This in turn, depends on the water balance for the site. The water balance can be written:

$$\text{Rainfall} + \text{leakage} - \text{runoff} - \text{evapotranspiration} = \text{infiltration into soil}$$

In this equation, leakage would include irrigation water for lawns and flower beds, leakage from plumbing and sewers, etc. Again, it is not easy to estimate an accurate water balance, but reasonable estimates for all of the terms can be obtained, if necessary by means of a Delphi process.

Once the rate of infiltration into the soil has been estimated, the time for heave to occur can be calculated from the availability of water and the time necessary to fill the air-filled pore space in the profile and the additional pore space created by swelling. Where the lateral extent of a structure subject to heave is of the same order as the depth of expansive material, water accumulates in the profile by ingress through the surface and then by lateral flow into the soil under the structure. The typical shape of the time versus heave curve in this case is ogival or *S*-shaped. Figure 8.29

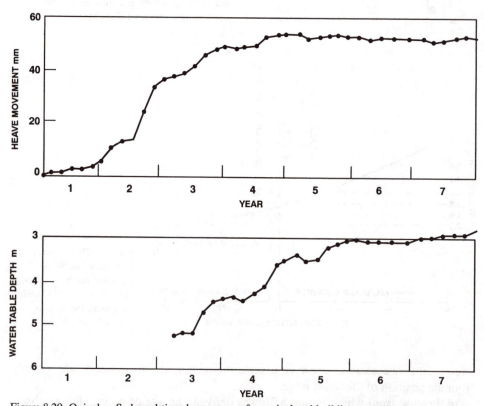

Figure 8.29. Ogival or S-shaped time-heave curve for an isolated building as moisture accumulates in the soil as a result of a change in land use (development as a housing estate). The effect of seasonal variations can be seen. The lower diagram shows the influence on the water table level.

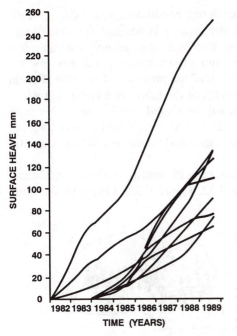

SURFACE HEAVE mm

TIME (YEARS)

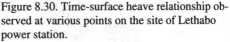

Figure 8.30. Time-surface heave relationship observed at various points on the site of Lethabo power station.

shows a typical ogival time-heave curve for a house built on a slab-on-grade where the depth of expansive soil was initially 5 m and the house measured 10 m × 6m. The figure also shows the rise of ground water level as water accumulated in the profile. When the area subject to heave has lateral dimensions that are large in comparison with the depth of expansive soil, the only way in which water can accumulate in the profile is by entering through the soil surface and migrating vertically downward. For this situation, it is assumed that the infiltration enters the soil as a sharply delineated wetting front and hence that heave proceeds from the top down. Because the more expansive soil is usually located near the top of the profile (see Fig. 8.27), the resulting time-heave curve is concave-up, and does not have the convex-up *S*-shape. Figure 8.30 shows a selection of time-heave curves observed for the profile referred to by Figures 8.26 and 8.27, that exhibit the concave-up shape.

8.4.3 *Collapse of residual soils*

The phenomenon of collapse settlement occurs in two types of residual soils. The first are the loess-like soils usually consisting of ancient wind blown soils that have been lightly cemented at the points of contact of the soil particles. The second type are unusually highly weathered and leached soils residual from acid rocks such as granites, that contain a large proportion of quartz. As a result of leaching and loss of mineral material, the residual soil becomes a silty or clayey sand with a high void ratio and an unstable collapsible grain structure. The following is a good example of the first type of collapsing soil:

Extensive areas of Southern Africa are covered with a blanket of wind-blown sands of Quaternary age. Because of changes of climate in the fairly recent past, these dune sands that were deposited under desert conditions, now exist in semi-deserts, e.g. as Kalahari grassland or even under savanna conditions. Because of present moister climate, the sands have partly weathered in situ and now contain a few percent of silt and clay. Originally deposited with a loose structure, this has now become a collapsible structure or fabric as a result of the presence of the clay (Knight 1961). Figure 8.31 shows Knight's classic sketch of the fabric of a collapsing wind-blown sand. As mentioned previously, leached weathered granites may also have collapsing properties. Figure 1.6 showed that increased annual rainfall over long periods results in greater leaching of weathered granite and therefore a higher void ratio and a greater tendency to be collapsible.

The characteristics of a collapsing soil have been illustrated by Figures 8.22a and 8.22b, and are further illustrated by Figure 8.32. Figure 8.32a illustrates the quite

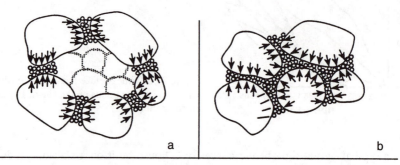

a   b

FLOCCULATED CLAY PARTICLES CONSOLIDATED BY CONCENTRATION OF PRESSURE

UNCONSOLIDATED FLOCCULATED CLAY PARTICLES

Figure 8.31. Knight's (1961) sketch of the mechanism of collapse of a collapsing sand. a) Loaded soil structure before soaking, b) Loaded soil structure after soaking.

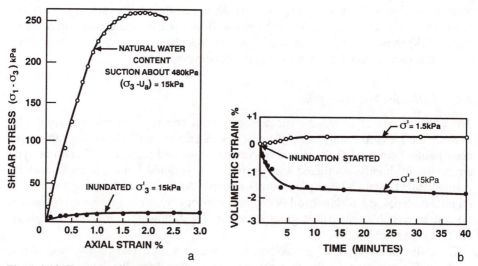

Figure 8.32. Tests on collapsing sand.

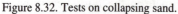

considerable strength (of about 275 kPa) of a collapsing soil at values of $(\sigma - u_a) = $ 15 kPa and a suction of 480 kPa. When the soil was wetted, the effective stress fell to 15 kPa and the strength was only 12 kPa. Figure 8.32b shows that the same soil will actually swell by a small amount if wetted under a very small confining stress (1.5 kPa). If it is wetted while carrying 10 times that stress (15 kPa), it suffers a relatively sudden collapse settlement of 2%. Hence a collapsing soil can have high strength and is relatively incompressible at low water contents (or high suctions) but loses strength and becomes compressible if the suction is reduced by an increasing water content.,

Although collapse cannot be predicted on the basis of effective stress considerations, collapsing soils behave as effective stress-controlled materials both before and after the collapse takes place (see Fig. 8.22). Depending on the water content of the soils, collapse may take place progressively, and not manifest as a sudden settlement at all. An example of this is shown in Figure 8.33 (Wagener 1985). If the soil of this example had been loaded at a time when its water content was high, it would have settled without collapse. If, on the other hand, it had been loaded at a low water content and was then subsequently wetted, it would have heaved slightly if loaded to less than 100 kPa, and collapsed if loaded beyond 200 kPa. There is thus a continuous spectrum between expansive and collapsing behaviour.

The amount of collapse settlement that occurs depends on the initial void ratio of the soil as well as the applied stresses. Under light foundation loads of 100 to 300 kPa, collapse settlements of up to 10% of the profile depth seem common, while settlements of up to 15% of profile depth have been reported.

Structures most likely to be affected by collapse are those founded at or near the surface, e.g. roads, housing and slab-on-grade floors to framed structures. Stiffened rafts have been used with apparent success, but relatively little research seems to

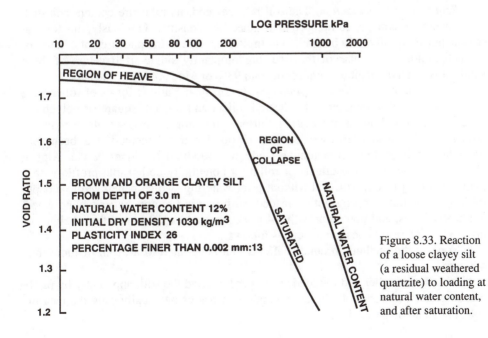

Figure 8.33. Reaction of a loose clayey silt (a residual weathered quartzite) to loading at natural water content, and after saturation.

have been done on the reasons for success of this type of foundation. To design the raft, it is usual to assume that collapse will occur in form of a 'soft spot' with a nominal diameter of up to 2 m, that can occur anywhere under the raft (Tromp 1985). Pile and pier solutions of various types have also been used (e.g. Schwartz & Yates 1980).

The amount of collapse that will occur is usually predicted on the basis of oedometer tests (e.g. Jennings & Knight 1975) that attempt to simulate the process of loading followed by wetting, that is likely to occur in the field. There appears to have been relatively little attempt to improve existing methods in recent years, even though the problem of collapse is wide-spread.

8.4.4 *In situ treatment of collapsing sands*

The most widely used in situ treatment for collapsing sands is to attempt to densify them by compaction, i.e. to pre-collapse them. One of the difficulties in adopting this approach is that not much is known about the depth to which a soil profile will collapse under load, and therefore the depth to which a potentially collapsible profile should be densified. Refuge is usually taken in the old rule of thumb for settlement, i.e. that collapse will extend to a depth of 1.5 times the least lateral dimension of a foundation. For roads and extensive hard-standing areas (e.g. container yards), it is the practice to specify compaction of the top 0.5 m to 90% of Mod AASHTO density and the next 0.5 m down, to 85% Mod AASHTO density. This has some basis in the observation that traffic loading on roads built over collapsing sands has been found to produce a densification to depths of just over 1 m (Knight & Dehlen 1963).

The practices used to achieve densification are limited only by the imagination of the designer or contractor. The following are few example of methods that have been tried:

– *Removal and compaction*: The soil is excavated, its moisture content adjusted, and it is then re-compacted in place to the specified densities. Obviously, this form of treatment can usually only be applied to limited areas, e.g. the plan area of a house or a small building. Because of the fine, predominantly single-sized nature of these sands, it is often difficult to compact beyond 93% of Mod AASHTO density.

– *Densification by rolling or pounding the surface*: Various forms of roller have been used, e.g. Vibrating smooth wheeled rollers and impact (square or polygonal) rollers, with or without prior watering. Surface pounding by drop-weight or dynamic compaction has also been used. Success reported with all methods has been very variable. An interesting comparative study was reported by Jones & van Alphen (1980) who compared the effects of rolling a potentially collapsible profile with a heavy vibrating roller, with and without prior watering (which penetrated to a depth of 300 to 500 mm), rolling with a loaded earth-moving motorscraper having wheel loads of 18 tons, and pounding with a 4 ton concrete block dropped from heights of up to 8 m. The results observed were as follows:

– None of the methods produced effects that penetrated to a depth of more than 400 mm.

– Initially the roller and the loaded scraper loosened the soil, apparently by partly breaking down its fabric. Only large numbers of passes eventually caused a measurable, but small densification.

– There was no difference in the behaviour of the watered and unwatered areas.

– Pounding loosened the soil and did not re-densify it.

Weston (1980) in a comprehensive review of the effects of rolling on collapsing sands for road construction has also reported mixed success in densifying collapsing sands in situ. Figure 8.34 reproduces some of Weston's data from this paper. Figure 8.34a shows the effect of compacting a section of road-bed with 20 to 30 passes of a heavy vibrating roller. Although some densification was achieved, there was relatively little change in the collapse settlements of the sand before and after rolling. Figure 8.34b shows results for a second section of the same road. Here, compaction was more successful in reducing the collapse potential of the sand. Weston concluded the following:

– The maximum relative compaction achieved by surface rolling is unlikely to exceed 90% of Mod AASHTO at depths of 0.5 to 1 m below the ground surface. The compactability is strongly influenced by moisture and clay content.

– There is no evidence to suggest that relative compactions in the top 0.5 m layer must be greater than 90% of Mod AASHTO for satisfactory road performance. In other words, the compaction results shown in Figure 8.34, although apparently disappointing. were satisfactory.

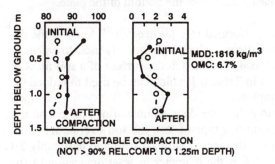

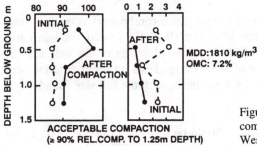

Figure 8.34. Results of rolling trials on the situ compaction of a collapsing Kalahari sand (after Weston 1980).

APPENDIX – DETAILS OF COMPRESSIBILITY TESTS AND SETTLEMENT CALCULATIONS

A8.1 DETAILS OF COMPRESSIBILITY TESTS

The following are some practical details relevant to performing in situ compressibility measurements:

A8.1.1 *The plate load test*

For a plate of breadth B, the plate load test gives a modulus of elasticity representative of the soil located within a depth of B to at most 2B beneath the plate. In general, the plate load test should be carried out at the level of the bottom of the footing or below this. Plate tests should only be conducted in the desiccated crust at the ground surface if the zone of actual footing influence lies primarily in that stratum. Frequently, plate load tests are performed at several depths in different strata. Before conducting a plate load test, the unsaturated soil beneath the plate should be soaked if it may be wetted during periods of prolonged wet weather. Also, careful soaking of the soil should be carried out if the test is intended to model deformation behaviour beneath the water table. Soaking should be continued for a sufficiently long period of time to wet the soil for a depth of at least 1.5B below the bottom of the plate.

Test pit. Usually the plate load test is performed in a pit excavated down to the desired level. However, elastic half-space theory, which is frequently used to calculate the modulus of elasticity, assumes the load is applied at the surface of a soil mass of wide lateral extent. The equation given in Table 8.1 which can be used to reduce the plate load test data, makes these assumptions.

If elastic half space theory is used to reduce the data, an unconfined plate load test should usually be performed. To perform this type of test, the soil overburden adjacent to the plate should be removed for a distance of at least 1.5B and preferably 2B away from the edge of the plate, where *B* is the diameter or least dimension of the plate. For a confined test, the soil surcharge should be left as close as possible to the plate. The effect of embedment should then be considered in calculating the modulus of elasticity, as discussed below.

Plate size and type. Both circular and square rigid plates can be used as well as cast-in-place concrete foundations. A desirable plate size (B) is 0.8 m with the minimum desirable size being about 0.3 m. The reliability of the plate test results increases as the plate size approaches the actual size of the footing. Tests on three different size plates can be used to help extrapolate settlements to those of the full size foundations (but see Fig. 8.12)

Deformation measurement. Plate settlement should be measured using at least 2 dial indicators (or other types of measurement devices such as LVDTs) placed diametrically opposite each other and at equal distances from the centre of the plate. Three displacement measurement devices are preferable to two. If three devices are used, space them 120° apart. The measurement devices should be capable of reading at least to the nearest 0.02 mm. These gauges should be tightly clamped to an independent reference beam supported on each side at least 1.5 m from the plate. (see Fig. 8.9).

Load application. Load is applied to the plate usually by an hydraulic jack. The jack can react against either a dead load platform or a portable lightweight truss held down by helix anchors or tension piles. It is desirable for a load cell, proving ring, or other measurement device to be used to measure the applied load accurately. The applied load must be carefully aligned perpendicular to the plate and applied through a ball and socket to minimize the detrimental effects of eccentricity.

To obtain good contact, a quick setting grout is placed between the plate and the soil. Plaster of paris (gypsum), either alone or mixed with a clean, fine sand or a quick setting commercial grout can be used to minimize bedding errors and help in levelling the plate. The grout bed should be as thin as practical and be allowed to cure before testing. Great care should be taken in levelling the plate and applying a perpendicular load at its centre.

Load test. The plate should in general be loaded to no more than the anticipated foundation design pressure to avoid excessive shear strains. Load should be applied in 5 equal increments allowing the settlement essentially to stop before applying the next increment. To determine the immediate settlement, take dial readings as soon as possible after applying the load for each increment. Also take readings at times of 0.5, 1, 2, 4, 8, 15, 30, 60 minutes, etc after load application. As the test progresses, plot plate settlement as a function of the logarithm of time. Leave the load on until the flat, straight line portion of the settlement-log time curve is reached.

After reaching the maximum load, unload in two increments. Application of at least a second load-unload cycle is recommended. A new load increment can be added in many residual soils after about 15 to 30 minutes or less, as elastic recovery is quick.

Primary consolidation settlement. For some saturated and/or clayey residual soils sufficient time may not be available, from a practical standpoint, to allow the delayed settlement to occur completely. As a practical expedient in testing these soils, the plate load test can be performed measuring only the instantaneous deformation. The modulus of elasticity calculated using the instantaneous deformation can then be corrected to consider, approximately, delayed settlement using one-dimensional consolidation test results. For each load increment in the one-dimensional consolidation test (or selected increments in the field plate load test), determine the immediate settlement as a percent of the total settlement (i.e. immediate plus primary). Do not include secondary compression in determining the total settlement. Plot the immediate settlement, expressed as a percent of total settlement, as a function of applied pressure. Now determine the average value of immediate settlement, expressed as a function of total settlement, over the pressure range applied by the foundation. Reduce the calculated modulus of elasticity to reflect the effect of delayed settlement.

Residual soils usually have instantaneous deformations greater than 60 to 70% of the total settlement. Assume, for example, the immediate modulus of elasticity is 9500 kPa, calculated from the plate load test results. Also assume that for this soil the average instantaneous deformation is 75% of the total deformation (excluding secondary compression). The corrected modulus of elasticity is then equal to 0.75×9500 kPa = 7125 kPa. This correction technique can also be used for correcting triaxial test values of E when the test has been performed too rapidly to allow delayed deformation to occur.

If the relative amount of immediate settlement compared to total settlement varies significantly over the range of applied load, the immediate settlement measured for each load increment can be corrected by dividing the measured immediate settlement by the ratio of immediate to total settlement for that increment as determined from the consolidation test.

Modulus of elasticity. The modulus of elasticity is calculated from the theory of elasticity. For a plate load test performed in the unconfined condition (i.e. no surcharge

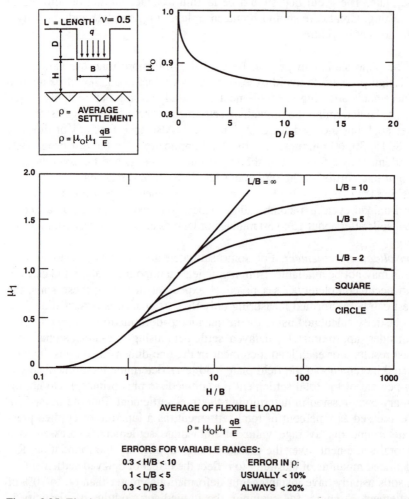

Figure 8.35. Elastic settlement calculation method for homogeneous soil: Rigid layer and foundation embedment (adapted from Christian & Carrier 1978).

near the sides of the plate), the equation given at the head of Table 8.1 can be used to calculate the modulus of elasticity. Use the appropriate influence factor I_f from Table 8.1 to account for the shape of the plate. For a confined test, use the elasticity approach summarized in Figure 8.35 which considers the depth of embedment used in the plate load test.

Soil disturbance. In performing a plate load test in a pit, the soil near the bottom of the plate undergoes the most disturbance because of loosening due to both the excavating and stress release. This disturbance can be minimized by excluding the deformation over a depth of about 0.5B beneath the plate. Good results have been reported in using this technique. The modulus of elasticity, however, must be backcalculated using appropriate theory and special instrumentation must be used to measure soil deformation at 0.5B depth below the plate.

A8.1.2 *The screw plate test*

Screw plate geometry. The screw plate geometry, illustrated in Figure 8.11 should have approximately the following ratios to minimize soil disturbance:

$$c/a = 0.125;\ b/c = 0.25;\ \text{and}\ t/a = 0.02$$

where: a = radius of screw plate, b = 1/2 of screw pitch, c = radius of loading shaft, t = screw plate thickness.

A tungsten tipped cutting edge on the screw plate is desirable to minimize wear caused by abrasion of the soil.

Test reactions. The test plate and loading shaft can react against a reaction frame or truss held down by dead load or helical tension anchors screwed into the ground, by a rotary drilling rig, or a heavily loaded vehicle.

Screw plate installation. Installation of the screw plate is accomplished by augering to within 400 mm of the desired test elevation. The screw plate is then advanced to the elevation of the test.

Use of a rotary drill rig and hollow stem auger sufficiently large to allow inserting the screw plate shaft into the hollow stem decreases the time required for performing a test. Using tension anchors to hold down the drill rig can substantially increase the reaction capacity. After augering the hole, the hollow stem drill rod is clamped. The inner loading shaft, which is attached to the screw plate by a hexagonal connection, is then advanced using the rotary feed on the drill rig.

Load test. The load test is conducted by applying load to the shaft which connects to the screw plate. A hydraulic jack is used to apply the load which is measured by either a calibrated proving ring or load cell. The combined deflection of the screw plate and elastic shortening of the loading shaft are measured using two dial gauges placed 180° apart. The dial gauges can easily be magnetically clamped to the loading shaft. Relative movement is measured between the loading shaft and a fixed reference beam which should be at least 3 m in length. Dial readings are taken from the time the drill rod is released from the chuck of the drill rig.

The screw plate loading procedure is to first apply a load equal to 20% of the estimated failure load and then remove it. This procedure reduces effects of soil disturbance and seats the plate. Next two load/unload cycles to the design pressure are applied, and, finally, the screw plate is loaded to failure and, if practical, unloaded.

For low plasticity soils, the compression rate should be about 0.25 to 1.25 mm per minute. In residual soils where primary consolidation is important, each load increment should if possible be left on until at least 70%, and preferably 100%, of primary settlement has occurred.

A8.1.3 *The pressuremeter test*

Hole preparation. The Menard type pressuremeter test should be performed in a smooth-sided hole. In residual soils, pushing a thin walled tube sampler ahead of a machine-augured bore hole has been found to work well. To obtain good results in residual soils, the cavity should be expanded by the pressuremeter to the stress level applied by the foundation at working load. Next release the stress exerted by the pressuremeter to a small seating value, and again increase the pressure back to the working stress level. The reload stress-volume change (or cavity radius) curve (see Fig. 8.13d) should be used in analysing the data. The load-unload-reload sequence takes out some effects of soil disturbance resulting from advancing the hole.

Equipment calibration. Pressuremeter test results and hence the modulus of elasticity determined from them are quite sensitive to equipment calibration and test correction effects. System calibration must take into consideration:

1. *System compliance.* System compliance includes both decrease in membrane thickness and also changes in the volume of pressure transmission tubes, etc. which both become increasingly important as the test pressure increases.

2. *Pressure effects.* Pressure loss must be accounted for due to membrane and sheath resistance to expansion. Also, differences in elevation between the pressuremeter and the pressure-measuring system must be considered if the pressure is measured at the surface.

All pressure gauges, transducers and deflection measurement devices must be accurately calibrated using reliable test standards.

Excellent books by Baguelin et al. (1978) and Mair & Wood (1987) are available which describe calibration and test procedures for various pressuremeters in detail as well as methods for calculating pressuremeter elastic moduli, shear strength parameters, and the lateral pressure ratio K_o. Empirical relations between pressuremeter modulus and standard penetration resistance or cone resistance have been successfully used in practice for the soils for which the correlations were developed.

A8.1.4 *Slow cycled triaxial test*

Stress state. A multi-stage test is performed by testing each specimen at three different confining pressures. Caution should be exercised to apply a stress state well below failure so as not to damage the usually delicate structure of the residual soil. Confining pressures should be selected which bound the stresses expected to exist beneath the completed structure. For many conventional building loadings, confining pressures of 35, 70 and 140 or 170 kPa are often suitable for use in the multi-stage triaxial test.

The deviator stress should be selected to cause a maximum shear stress in the specimen no greater than about 15% of the maximum failure shear stress. To obtain an acceptable maximum deviator stress, estimate using past experience for the material to be tested.

After selecting appropriate shear strength parameters, plot the failure envelope using the same scale for the normal stress (abscissa) and shear stress (ordinate) axes. For each confining pressure to be used in the multi-stage test, draw a failure circle on the plot limiting the maximum shear stress to no more than about 30% of the estimated failure stress. Table 8.3 illustrates typical stress states suitable for residual silty sands and sandy silts.

Cycled test. At least two complete load-unload cycles of testing should be carried out for each of the three confining pressures used in the test. The specimen should be tested in a saturated condition if during the structure's design life the soil may reach a high degree of saturation. A consolidated, drained triaxial test is used so as to allow as much consolidation as possible to occur during the test.

After initially applying each confining pressure, measure axial specimen displacement as a function of time to determine when primary specimen consolidation is complete. Upon reaching the end of primary consolidation, begin applying the deviator stress load-unload cycles. Preferably a maximum axial straining rate of about 0.2%/min should be used in the test. For low permeability residual soils having important delayed settlements, the results should be corrected if the test is not performed sufficiently slowly to measure time delay effects. A practical method for correcting the modulus of elasticity for time delay effects using consolidation test results was previously described in the plate load testing section (Section 8.1.1).

Modulus of elasticity. For each confining pressure, determine the modulus of elasticity from each unload cycle. Usually the modulus of elasticity for each subsequent load-unload cycle is greater than for the previous cycle. The rate of increase, however, becomes less with increasing numbers of load-unload cycles. The modulus of elasticity is equal to the absolute value of the change in deviator stress during unloading divided by the smaller of the change in height of the specimen or the change in axial distance between displacement measurement points on the specimen, expressed as an axial strain.

Table 8.3. Typical stress states for slow cyclic triaxial testing of silty sand and sandy silt residual soils.

Applied Stresses (kPa)	Soil consistency					
	Stiff ($N = 8$ to 15)		Firm ($N = 5$ to 8)		Soft ($N = 3$ to 4)	
σ_3	$\sigma_1 - \sigma_3$	σ_1	$\sigma_1 - \sigma_3$	σ_1	$\sigma_1 - \sigma_3$	σ_1
35	40	75	30	65	20	55
70	50	120	40	110	30	100
140	100	240	70	210	60	200

Usually the modulus of elasticity obtained from the last unload cycle is used in design. To aid in selecting design values of the modulus of elasticity, plot all of the measured elastic moduli for a particular site on log-log graph paper. The modulus of elasticity is plotted on the vertical axis and confining pressure on the horizontal axis. Curve fit (by hand or computer) the data points obtained for each specimen (i.e. the modulus of elasticity measured at the three confining pressures). Label each curve (which represents the results for one specimen) including a complete soil description of the specimen, depth of sampling, and the observed value of standard penetration resistance, cone tip resistance, etc.

A8.2 SETTLEMENT PREDICTION CALCULATIONS

The prediction of settlement is a critical aspect of foundation design. Settlements of structures on residual soils are usually determined using one of the following two techniques:

1. One-dimensional consolidation test results and stress distribution theory.

2. Elastic theory which uses a modulus of elasticity measured in the field or laboratory or determined from empirical correlations with field measurements.

Accurate settlement predictions are hard to perform for structures founded on residual soils. Soil disturbance effects together with testing errors in both the laboratory and field are important factors which help to account for actual settlements frequently being less than measured values by 30 to 50%.

Considerable progress has been made in recent years towards understanding the behaviour of residual soils. The use in settlement prediction methods of elastic moduli measured in situ has given improved results. Also, the measurement of axial deformation within the middle portion of a triaxial test specimen subjected to a small strain level has been found to give more realistic results (i.e. higher elastic moduli) than using conventional methods for measuring overall specimen deflection. Nevertheless, the available settlement prediction approaches have not proved entirely successful for all soil and loading conditions. Local experience, developed by comparing predicted and measured settlements, is still needed to verify settlement calculation results.

A8.2.1 *Selection of settlement prediction methods*

The analytical approach for predicting the settlement of residual soils must have a sound theoretical basis, be able to handle varying soil and foundation conditions, and be suitable for use with a variety of laboratory and field testing methods.

To satisfy these widely varying needs, an elasticity-based approach has been selected. For shallow foundations this includes the strain influence diagram method originally proposed for sands by Schmertmann (1970). The strain influence diagram method has proven to be well-suited for predicting the settlement of shallow foundations on most residual soils. Elasticity-based methods are not suitable for predicting settlements of collapsible or swelling soils. This method should not be used with high plasticity residual soils which exhibit more than 40 to 50% delayed settlement unless local experience shows it to be a valid approach.

A8.2.2 *Modulus of elasticity*

A reliable value of the modulus of elasticity of the soil strata contributing to settlement must be evaluated for use in elasticity-based settlement approaches. The method used for evaluating the modulus of elasticity depends upon the sophistication of the project and availability of field and laboratory testing equipment. Methods that can be used to estimate elastic moduli include:

In situ tests: The plate load test, screw plate test, or more sophisticated tests such as the pressuremeter or dilatometer test can all be used to determine the modulus of elasticity directly. In general these methods are preferable to laboratory tests. The plate load test perhaps gives the best representation of the in situ modulus but is labour intensive to perform.

Laboratory tests: The triaxial and one-dimensional oedometer (consolidometer) tests have also been quite popular in many parts of the world.

Empirical correlations: Empirical correlations can be used to relate the modulus of elasticity to empirical indicators of stiffness such as standard penetration resistance or cone tip resistance. The use of empirical correlations can be employed to advantage for routine projects or when very limited field and/or laboratory capability is available. The empirical correlation approach should only be employed for geological units for which past settlement experience is available and which can be used to verify the accuracy of the correlations.

The application of these methods was discussed earlier for estimating the modulus of elasticity to use in predicting settlement of residual soils.

A8.2.3 *Strain influence diagram method*

The strain influence diagrams used in the method are based on vertical strains measured in model studies and calculated using the finite element method. Vertical strain has been observed in a number of laboratory and large scale field studies to die out more rapidly than indicated by Boussinesq stress distribution theory for a homogeneous soil. Both measurements made on model foundations and finite element theory also show that the maximum strain does not occur immediately beneath a rigid foundation as implied by the usually used stress distribution-based settlement prediction methods. Instead, the maximum strain is developed between about 0.5 and 1 times the diameter or width of the foundation as shown in Figure 8.36. Therefore, the strain influence diagram approach agrees better with observed fundamental foundation behaviour than do the commonly used stress distribution based settlement methods.

The strain-influence diagram method is quite versatile and easy to use. The vertical settlement of a foundation is equal to the integral of the vertical strain over the depth of settlement influence beneath the point at which settlement is desired. Numerical integration of strain with depth is easily performed by first dividing the depth influenced by the loading into sublayers of thickness Δz. The total foundation settlement is then the summation of average vertical strain in each sublayer multiplied by its thickness Δz.

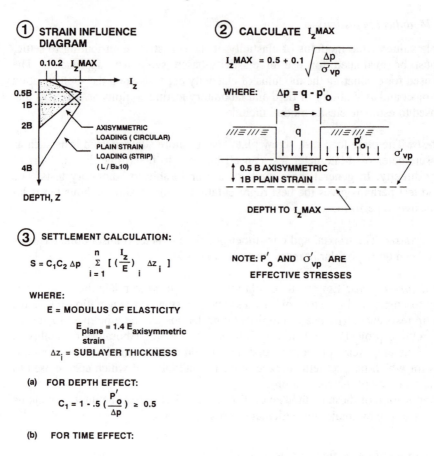

Figure 8.36. Summary of strain influence diagram settlement calculation method.

The strain influence diagrams used in calculating foundation settlement for rigid circular and rigid infinitely long foundation on a homogeneous soil are shown in Figure 8.36. To perform a settlement analysis using the strain influence diagram method, first divide the strata beneath the foundation into n convenient sublayers of thickness Δz_i. Sublayer boundaries are usually placed at the break in the strain influence diagram and where changes in the modulus of elasticity occur such as at boundaries between different strata. Typically 2 to 4 sublayers are used in a settlement analysis. The foundation settlement (*S*) is then calculated using the following generalized equation:

$$S = C_1 C_2 \Delta p \sum_{i=1}^{i=n} (I_z / E)_i \Delta z_i \qquad (A8.1)$$

where S = total foundation settlement, n = total number of sublayers, C_1, C_2 = constants accounting for foundation depth and time effects, respectively, Δp = net increase in effective pressure applied by the foundation, Δz_i = variable increment of depth whose sum equals the depth of the strain influence diagram, I_z = average strain influence factor over the increment of depth Δz_i, E = average modulus of elasticity of the soil over the corresponding increment of depth Δz_i.

The correction factor C_1, used in Equation (A8.1) reduces the calculated settlement to account for the beneficial effect of foundation embedment:

$$C_1 = 1 - 0.5(p'_o/\Delta p) \tag{A8.2}$$

where p'_o = initial effective overburden pressure (refer to Fig. 8.36), Δp = net increase in effective foundation pressure.

The correction factor C_1 should not exceed 0.5 for deeply embedded foundations. For foundations embedded less than 0.7 m, the use of $C_1 = 1.9$ is suggested.

The C_2 factor in Equation (A8.1) accounts for secondary compression of the soil. Secondary compression occurs over a period of time primarily after excess pore pressures caused by the applied loading dissipate. The constant C_2 is calculated from the expression:

$$C_2 = 1 + 0.2 \log (10t) \tag{A8.3}$$

where t is the time in years after the load is applied.

The rate at which secondary compression occurs is related to the compressibility of the soil. As a result, density, soil structure, mineral composition of the soil, including mica content, and other variables all influence the rate of secondary settlement. Although Equation (A8.3) does not directly consider these factors, this approach appears to give fair estimates of secondary compression effects based on somewhat limited verification for residual soils.

Following the strain influence diagram method, the settlement is calculated using the following general method which involves integrating the vertical strain (ε_v) over the depth of settlement influence zMAX below the bottom of the foundation:

$$S = \sum_{z=0}^{z=z\text{MAX}} \varepsilon_v dz \tag{A8.4}$$

The vertical strain ε_v, occurring at a given depth z beneath the loaded area can be expressed as

$$\varepsilon_z = \Delta p \, (I_z/E_s) \tag{A8.5}$$

where all the terms have been previously defined. Inserting Equation (A8.5) into Equation (A8.4) gives:

$$S = \Delta p \sum_{z=0}^{z=z\text{MAX}} (I_z / E_s) \Delta z \tag{A8.6}$$

where no new terms have been introduced. The increase in net pressure applied at the base of the foundation, Δp, was taken outside of the summation since it is not a function of depth. Now, numerically integrate Equation (A8.6) over the depth of strain influence below the base of the foundation. Then modify the calculated settlement by the correction factor C_1 for depth effect and C_2 for secondary compression. This manipulation gives the generalized Equation (A8.1).

A8.2.4 *Settlement prediction for special conditions*

Circular and long strip footings on homogeneous, deep layer. Figure 8.36 gives strain influence diagrams for rigid circular and strip footings. These footings rest on a deep layer having an approximately constant modulus of elasticity (i.e. stiffness) with depth. To avoid tilt, the centroid of the applied loading must coincide with the centroid of the foundation. For most accurate results, the depth to rock must be greater than about 3B beneath the bottom of circular footings and 6B beneath the bottom of very long footings. The ground surface should be reasonably level.

The elasticity-based solutions given in Figure 8.37 and Table 8.1 can also be used to calculate settlement for the conditions described above. Figure 8.37 and Table 8.1 are particularly useful for uniform soil conditions, as discussed subsequently.

Adjacent footings. The additive effect of adjacent footings can be estimated using superposition and the solutions given in Table 8.1 and Figure 8.37. The use of super-position is illustrated in Example 6 (below).

The settlement caused by adjacent footings can be estimated using Figure 8.37. This figure is for a rigid, circular footing resting on a deep, homogeneous layer. Figure 8.37 shows a cross section taken through the centre of the footing and the soil

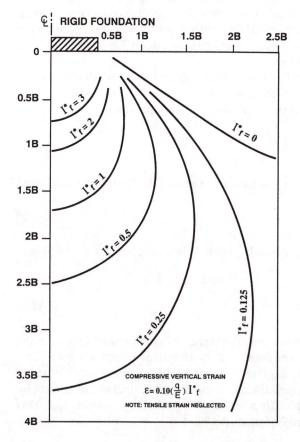

Figure 8.37. Contours of equal vertical strain influence factor I_f^* for rigid circular foundation: Deep homogenious soil mass.

underlying it. All dimensions are in terms of the foundation diameter. Superimposed on this cross section are contours of equal vertical strain influence factor I_f^* as a function of depth. Now determine the settlement by using the strain influence factors and the slightly modified form of Equation (A8.1).

$$S = 0.10\, C_1 C_2 \Delta p \sum_{i=1}^{i=n} (I_f^* / E)_i \cdot \Delta z_i \qquad (A8.7)$$

All terms in the above equation have been previously defined and discussed.

As shown on the contour influence diagram given in Figure 8.37, vertical strain caused by the applied loading dies out quite quickly with increasing lateral distance from the footing. For many problems, the effect of adjacent footings, when greater than 2B away, can therefore be neglected.

Great depth. Table 8.1 can also be used for square, circular and rectangular footings resting near the surface of a deep layer of homogeneous soil. The generalized elasticity equation used to calculate settlement is given at the top of the table.

Rectangular foundations: Generalized strain influence diagrams. Consider the determination of settlement of a rigid, rectangular foundation of varying shape resting on a deep homogeneous stratum. Settlement can be estimated using the strain influence diagram method for rectangular foundations (Fig. 8.38). As the length to width ratio, L/B, of the foundation becomes greater, the following quantities associated with the strain influence diagram increase: (a) the dimensionless maximum depth of the influence diagram, zMAX /B, (b) the dimensionless depth below the bottom of the footing to the maximum strain influence factor, zPK/B, and (c) the value of the strain influence factor, I_z° at the bottom of the footing.

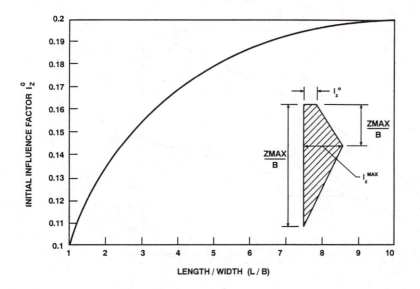

Figure 8.38. Strain influence factor I_z° at base of rectangular foundation as a function of L/B.

For a given value of the length to width ratio, L/B, of the foundation, determine from Figure 8.38 the value of the strain influence factor $I_z°$ at the bottom of the foundation. This figure also shows the notation used for the influence diagram variables which are a function of L/B. The maximum depth, zMAX /B, of the diagram and depth to the peak value (zPK/B) are obtained from Figure 8.39. Multiply the dimensionless values obtained from this figure by the foundation width (i.e. the minimum dimension of the rectangular foundation) to obtain the actual value of the variable. The settlement calculations are carried out using Equation (A8.1) in the same manner as previously discussed.

Flexible circular, square and rectangular foundations, homogeneous deep strata. A flexible foundation is one that has a small stiffness when subjected to bending. A fill can be considered as a flexible loading as well as a thin, lightly reinforced concrete mat foundation. A centrally loaded, flexible foundation has a dish-shaped settlement profile with the greatest settlement occurring in the centre and the least settlement at the edges.

The solutions given in Table 8.1 can be used directly to calculate the settlement of flexible foundations resting on a deep layer of homogeneous soil. The strain influence diagram method can also be used for flexible foundations. First calculate the settlement for a rigid foundation having the correct shape using the strain influence diagram method. Then correct the results by multiplying the calculated rigid foundation settlement by the correction factors given in parentheses in Table 8.1. For either method of settlement calculation, superposition can be used to obtain settlements at locations on the foundation for which influence factors are not given in Table 8.1.

Circular rigid foundation – Increasing stiffness with depth. The stiffness of residual soils frequently increases with depth. The increase in stiffness is at least partly due to the degree of weathering becoming less and confining stress greater with increasing

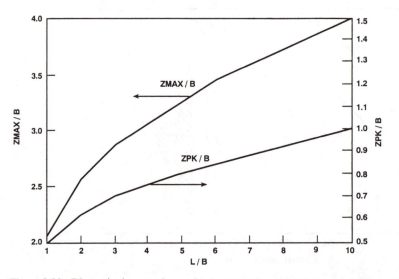

Figure 8.39. Dimensionless maximum depth zMAX /B and depth to peak influence factor zPK/B, as a function of L/B.

depth. For this type of soil stiffness profile, the vertical strain developed beneath a footing dies out with increasing depth more rapidly than for a homogeneous soil.

To approximate this condition, first consider a rigid, circular footing of diameter B underlain by a deep soil stratum whose stiffness increases linearly with depth. The elasticity solutions for these conditions given in Table A8.1 or the strain influence diagram approach for a homogeneous soil (Fig. 8.36) can still be used, provided the calculated settlement is corrected using Figure 8.40. First calculate the settlement for a circular footing resting on a deep homogeneous stratum. Use in these calculations the soil stiffness at a depth beneath the foundation equal to its diameter B. Then multiply this settlement by the appropriate correction factor C_3 given in Figure 8.40.

Figure 8.41, which is similar to Figure 8.40 is used to calculate the settlement when a rigid layer is at a depth of 2B beneath the foundation. Once again, the stiffness increases linearly with depth below the circular, rigid foundation. The modulus of elasticity at a depth of B beneath the foundation is used. This calculated settlement is then multiplied by the appropriate correction factor C_4 given in Figure 8.41.

A8.2.5 *Menard method for calculating settlement of shallow foundations*

The prediction of settlement is made via a semi-empirical equation:

$$\rho = \frac{q - \sigma'_{vo}}{9 E_M}\left[2 B_o \left(\lambda_d \frac{B}{B_o} \right)^\alpha + \alpha \lambda_c B \right] \tag{A8.8}$$

In this equation the first term in [ ] is said to account for settlement caused by shear distortion, while the second term accounts for compressional settlement. q = gross average bearing stress, σ'_{vo} = effective overburden stress at founding level, B_o = a

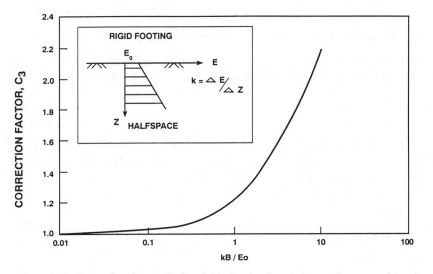

Figure 8.40. Correction factor C_3 for rigid circular foundation resting on a deep, homogeneous layer; modulus of elasticity increases linearly with depth.

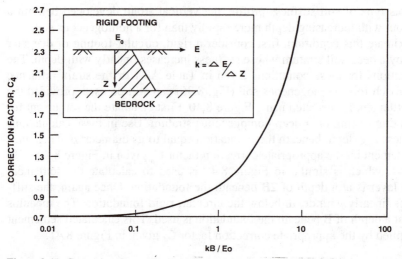

Figure 8.41. Correction factor $\acute{C}_4$ for rigid circular foundation resting on layer 2B in depth; modulus of elasticity increases linearly with depth.

reference width, usually 600 mm, B = width or diameter of footing where $B \geq B_o$, α = a rheological factor that depends on soil type and the ratio E_M/p_l^*, p_l^* = net limiting pressure = $p_l - \sigma'_{w}$, λ_d, λ_c = shape factors depending on the length/width ratio L/B) of the foundation.

Equation (A8.8) is said to apply in cases where the depth of founding D is at least equal to the footing width B, i.e. $D \geq B$. Where $D < B$ the following correction factors are applied (Table 8.4):

The values of for silt, sand and sand-gravel mixtures as well as the shape factors are tabulated in Tables 8.5 and 8.6.

A8.2.6 *Sellgren's method for predicting the settlement of piles*

Sellgren (1981) suggested that the load displacement curves of single driven piles should be taken to be in the form of a hyperbola as follows (Fig. 8.42)

$$S = \frac{aP}{1 - bP} \tag{A8.9}$$

where S = pile head settlement, P = applied vertical load, a = constant, α = slope of initial part of hyperbola, $a = \tan \alpha$ in Figure 8.42, b = constant, the inverse of which is the asymptote to the curve, i.e. $P_f = 1/b$.

Equation (A8.9) thus becomes

$$S = \frac{aP}{\left(1 - \dfrac{P}{P_f}\right)} \tag{A8.10}$$

where P_f = ultimate vertical load.

Table 8.4. Menard's correction factors for foundation embedment.

D/B	Correction factor
0	1.2
0.5	1.1
1	1.0
> 1	1.0

Table 8.5. Values of rheological factor α.

Soil type	Silt		Sand		Sand and gravel	
	E_M/pl^*	α	E_M/pl^*	α	E_M/pl^*	α
Over-consolidated	> 14	2/3	> 12	1/2	> 10	1/3
Normally consolidated	8-14	1/2	7-12	1/3	6-10	1/4
Weathered and/or remoulded		1/2		1/3		1/4

Table 8.6. The shape factors λ_d and λ_c.

L/B_o	1		2	3	5	20
	Circle	Square				
λ_d	1	1.12	1.53	1.78	2.14	2.65
λ_c	1	1.10	1.20	1.30	1.40	1.50

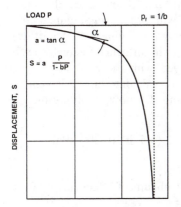

Figure 8.42. Load displacement relationship proposed by Sellgren.

From the results of many pile tests, it has been found that factor '*a*' can be determined by the expression given as Equation (A8.11):

$$a = \frac{4}{\pi D_p} \cdot \frac{1+\dfrac{\beta}{(\theta E_p D_p)\tanh\theta l_p}}{\beta + \theta E_p D_p \tanh\theta l_p} \tag{A8.11}$$

where D_p = pile diameter, l_p = pile length, $B = 6E_m/(1 + v)$, $\theta = \sqrt{4B/E_p D_p}$, $B = 4.17$ E_m for $v = 0.3$, E_m = pressuremeter modulus, E_p = Young's modulus for pile material.

For piles with square cross section, the term $1/B$ is substituted for $4/\pi D_p$ in Equation (A8.11) where B is the width of the pile side.

A8.3 SETTLEMENT CALCULATIONS – EXAMPLES

Example 1: Linearly varying modulus. Consider a rigid, 3 m diameter foundation that is 0.7 m thick. The bottom of the foundation is 0.7 m below the surface. The pressure exerted on the soil by the column (including the weight of the footing) is 155.6 kPa. The stiffness of the soil, as defined by the modulus of elasticity, increases approximately linearly with depth. The modulus of elasticity is estimated to be 5 MPa at the bottom of the footing and 15 MPa at a depth of 6 m below the foundation. The soil is partially saturated and weighs 19 kN/m^3. Estimate the settlement of the rigid, circular footing using the strain influence diagram approach.

$$\Delta p = 155.6 \; kPa - 0.7 \; m \; (19 \; kN \, / \, m^3) = 142.3 \; kPa$$

$$\sigma'_{vp} = [0.7 \; m + 0.5 \; (3 \; m)] \, 19 \; kN \, / \, m^3 = 41.8 \; kPa$$

The maximum strain influence factor I_z^{max} obtained using the equation given in Figure 8.36 is

$$I_z^{max} = 0.5 + 0.1 \sqrt{142.3 / 41.8} = 0.68$$

Now use Equation (A8.1) and the modulus of elasticity at a depth of 1B below the bottom of the foundation to calculate settlement ($E = 10$ MPa). Then correct the foundation settlement to account for the linear variation of E with depth using Figure 8.40.

The scale plot of the I_z strain influence diagram shown in Figure 8.43 can, if desired, be used to determine the average values of the strain influence factor for each sublayer. In this example, two sublayers are employed with the boundary being placed at the break in the strain influence diagram. Now calculate the settlement of the rigid circular footing using Equation (A8.1) without, for now, applying the C_1 and C_2 correction factors (i.e. let $C_1 = C_2 = 1$):

$$S_1 = 142.3 \; kPa \; [0.392 \; (1.5 \; m)/10,000 \; kPa + 0.343 \; (4.5 \; m)/10,000 \; kPa]$$
$$S_1 = 30.3 \; mm$$

Since the foundation is relatively shallow, neglect the embedment effect (i.e. let $C_1 = 1$). Correct for secondary compression (Eq. A8.3) using $t = 3$ years.

$$C_2 = 1 + 0.2 \log (3/0.1) = 1.3$$
$$S_2 = C_2 S_1 = 1.3 \; (30.3 \; mm) = 39.4 \; mm$$

The calculated foundation settlement of 39.4 mm, which is valid for a uniform modulus of elasticity of 10 MPa, must be corrected to consider the increasing modulus of elasticity with depth. Since the soil stratum is deep, use Figure 8.40 to obtain the correction C_3. In Figure 8.40 the constant $k = \Delta E / \Delta z = (10,000 \; kPa)/6 \; m = 1667 \; kN/m^3$ and $kB/E_o = (1666.7 \; kN/m^3 \, (3 \; m)/5000 \; kN/m^3 = 1.0$. The variable E_o is the modulus of elasticity at the bottom of the foundation (refer to Fig. 8.40). From Figure 8.40 the correction factor $C_3 = 1.22$ which gives a corrected settlement of

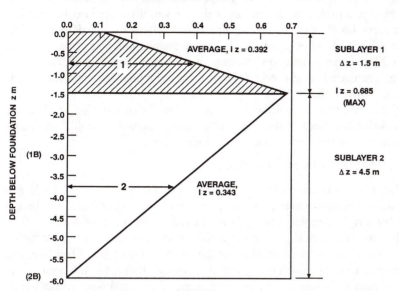

Figure 8.43. Strain influence factor diagram for Example 1.

S_3=1.22 (39.4 mm) = 48.1 mm which is the desired estimate of foundation settlement. The estimated settlement should be reported to the nearest 5 mm, as 50 mm.

Example 2: Elasticity equations. An alternate method for calculating the settlement for the rigid circular foundation problem given in Example 1 is to use the elasticity equation given at the top of Table 8.1. This equation is valid for deep homogeneous strata:

$$S = [C_2 \cdot C_3 \Delta p B (1 - v^2) I_w] / E \qquad (A8.12)$$

The constants C_2 and C_3 in the above equation, which were previously discussed, were added to correct the elasticity equation given in Table 8.1 for secondary compression and a linearly increasing modulus of elasticity with depth. To neglect embedment effects, once again let $C_1 = 1.0$.

Now let Poisson's ratio $v = 0.35$ and use the constants $C_2 = 1.3$ and $C_3 = 1.22$ which are the same as for Example 1. The influence factor $I_f = 0.785$, obtained from Table 8.1 for a rigid circular foundation, is used in Equation (A8.12) giving

$$S = [(1.3) (1.22) (142.3 \text{ kPa}) (3 \text{ m}) (1-0.35^2) 0.785]/10\ 000 \text{ kPa} = 46.6 \text{ mm}$$

The calculated foundation settlement of 46.6 mm obtained using the elasticity equation given in Table 8.1 compares favourably with the value of 48.1 mm calculated using the strain influence diagram approach. The elasticity equation approach for straightforward settlement problems is easier to apply than the strain influence diagram method. The important advantage of the strain-influence diagram method, however, is in applying it as an approximation to problems where the soil stiffness varies in an irregular manner as in Example 5.

Example 3: Square foundation. Assume the problem described in Example 1 involves a 3 m by 3 m square foundation rather than a 3 m diameter circular foundation. All other variables are the same as for Example 1. Using the elasticity equation given in Table 8.1 with $I_f = 0.815$ for a rigid square foundation, and as before $C_2 = 1.3$ and $C_3 = 1.22$ gives an estimated settlement of 48.4 mm using $E = 10,000$ kPa at a depth of 1 m. The equations given in Table 8.1 are for a homogeneous soil which has an E constant with depth.

A good estimate of settlement of the rigid square foundation can also be obtained using the strain influence diagram method. Correct the settlement ($S = 48.1$ mm) calculated in Example 1 for the rigid 3 m diameter foundation as follows:

$$S = (0.815/0.785)\ 48.1\ \text{mm} = 49.9\ \text{mm}$$

The 0.815/0.785 correction used above is the influence factor I_f given in Table 8.1 for a rigid square foundation having a width $B = 3$ m ($I_f = 0.815$) divided by the influence factor for a rigid circular footing (0.785) having a diameter of $B = 3$ m ($I_f = 0.785$). For the two foundations, all other variables are the same in the elasticity equation, used to calculate settlement, given at the bottom of the table. Therefore, the ratio of influence factors is equivalent to multiplying the settlement of the rigid circular footing calculated in Example 1 by the settlement of the rigid square foundation divided by that of the rigid circular foundation for a homogeneous isotropic solid. This manipulation gives the correct settlement for a homogeneous soil and an approximate settlement for the case of a linearly increasing modulus with depth.

Example 4: Strain diagram method with linearly varying modulus. To illustrate the versatility of the strain influence diagram method, once again solve the rigid circular foundation problem given in Example 1. Use the actual linear variation with depth of the modulus of elasticity. The strain influence diagram for a rigid, circular foundation of $B = 3$ m is 6 m deep (i.e. 2B) below the bottom of the footing as shown in Figure 8.36. Divide this 6 m depth into 3 sublayers ($\Delta z = 1.5$ m, 2 m and 2.5 m), and read the influence factors ($I_f = 0.392, 0.533$ and 0.190) from the strain influence diagram given in Figure 8.44 which has been plotted to scale. The average modulus of elasticity at the centre of each sublayer is 6250, 9167 and 12,917 kPa. Now substitute the appropriate values into Equation (A8.1) and letting for now the correction factors $C_1 = C_2 = 1.0$ gives:

$$S = 142.3\ \text{kPa}\ \frac{[0.392\ (1.5\ \text{m})}{6250\ \text{kPa}} + \frac{0.533\ (2\ \text{m})}{9167\ \text{kPa}} + \frac{0.190\ (2.5\ \text{m})]}{12,917\ \text{kPa}} = 35.1\ \text{mm}$$

Following Example 1, neglect the effect of foundation depth (i.e. let $C_1 = 1$) and correct for secondary compression ($C_2 = 1.3$) which gives

$$S = 1.3\ (35.1\ \text{mm}) = 45.6\ \text{mm}$$

The computed settlement is slightly less than the theoretically more accurate estimate of 48.1 mm determined in Example 1. From a practical viewpoint, however, the difference in answers is not significant.

The generalized strain influence diagram method, as illustrated by this example, can be used to best advantage when the soil stiffness beneath a foundation varies in

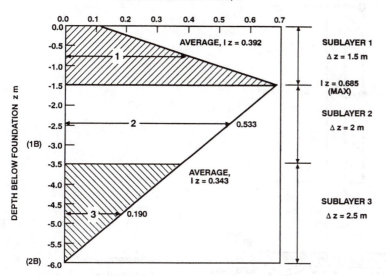

Figure 8.44. Strain influence factor (I_z) diagram for Example 2: E_s increasing linearly with depth.

an irregular manner for which more accurate solutions are not available. Refer to the next example to further illustrate this approach.

Example 5: Flexible foundation. Estimate the settlement of the flexible foundation shown in Figure 8.45 at points A, B and C. The rectangular foundation which is 4 m wide and 20 m long (L/B = 5), is loaded by a uniform pressure of q = 40 kPa which includes the weight of the foundation. As also shown in Figure 8.45, a soft layer 3.24 m thick (E = 3000 kPa) overlies a 20 m thick stiff layer (E = 6000 kPa). The groundwater table is located at the surface, and the bottom of the foundation is embedded 2 m. The modulus of elasticity of each layer was measured in the laboratory using slowly cycled, drained triaxial tests.

The modulus of elasticity for an infinitely long foundation can be taken as 1.4 times that obtained for an axisymmetric loading (refer to Fig. 8.36). An axisymmetric loading is applied, for example, in both the triaxial test and the field plate load test which uses a circular plate or, approximately, a square plate. Assume that an infinitely long foundation exists when L/B > 10. Now interpret linearly between L/B = 1, for which correction of the modulus of elasticity is not required, and L/B = 10 where the corrected modulus (E') is equal to E'_s = 1.4 E. The corrected modulus for this problem (L/B = 5) then becomes E' = 1.2 E.

First solve the problem of a rigid, rectangular foundation, and then correct, approximately, the results to a flexible foundation using Table 8.1. For L/B = 20m/4 m = 5, the initial influence factor $I°_z$ = 0.179 (Fig. 8.38). From Figure 8.39, ZMAX/B = 3.27 (read the left vertical axis) and hence ZMAX = 3.27 (4 m) = 13.08 m. Also from Figure 8.39, ZPK/B = 0.810 and hence ZPK = 0.810 (4 m) = 3.24 m. The resulting strain influence diagram for the rigid, rectangular foundation is summarized in Figure 8.45.

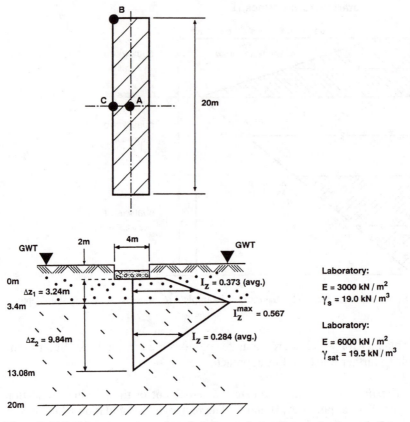

Figure 8.45. Problem geometry and strain influence factor diagram for example 5.

Now determine the maximum strain influence factor I_z^{max}. The net increase in effective pressure applied to the soil is: $\Delta p = q - p_o = 40$ kPa $- 2$ m [19.0 $-$ 10] kPa $= 21.6$ kPa. (Note that in this calculation the buoyant soil unit weight is used below the ground water table). The level at which to calculate σ_{vp} is equal to the foundation embedment depth of 2 m added to ZPK (3.24 m):

$$\sigma_{vp} = (2 \text{ m} + 3.24 \text{ m}) (19.0 - 10) \text{ kPa} = 48.2 \text{ kPa}$$
$$I_z^{max} = 0.5 + 0.1 (21.6 \text{ kPa}/48.2 \text{ kPa})^{1/2} = 0.567$$

Upon correcting the modulus of elasticity measured in the laboratory by 1.2, Equation (A8.1) and once again letting $C_1 = C_2 = 1$, gives

$$S_1 = 21.6 \text{ kPa} \left[\frac{0.373 \, (3.24 \text{ m})}{3000 \text{ kPa} \, (1.2)} + \frac{0.284 \, (9.84 \text{ m})}{6000 \text{ kPa} \, (1.2)} \right] = 15.6 \text{ mm}$$

Since the foundation is relatively deep (2 m), calculate the embedment correction factor C_1:

$$C_1 = 1 - 0.5\,(p_o' \,/\, \Delta p) = 1 - 0.5\frac{(18.4 \text{ kPa})}{21.6 \text{ kPa}} = 0.574$$

Since 0.574 is greater than the limiting (minimum) value of $C_1 = 0.5$, use 0.574 in reducing the settlement. For an elapsed time of 3 years, C_2 was previously found to be 1.3. The settlement for a rigid, rectangular foundation having L/B = 5 then becomes

$$S_2 = C_1 C_2 S_1 = (0.574)\,(1.3)\,15.6 \text{ mm} = 11.6 \text{ mm}$$

Now correct the 11.6 mm settlement calculated for the rigid foundation to approximate the settlement that would occur beneath a flexible foundation. To make these corrections, use the factors given in parentheses in Table 8.1 for L/B = 5.

> Foundation Centre (Point A): $S = 1.35\,(11.6 \text{ mm}) = 15.7 \text{ mm}$
> Foundation Centre (Point A): $S = 0.68\,(11.6 \text{ mm}) = 7.9 \text{ mm}$
> Foundation Edge (Point C): $S = 1.08\,(11.6 \text{ mm}) = 12.5 \text{ mm}$

As expected, the settlement at the centre of the flexible foundation is greater than for the rigid foundation. At the corner the settlement of the flexible foundation is less than for the rigid foundation.

Example 6: Flexible foundation – superposition. Consider a 5 m by 25 m bin filled 4 m high with coal weighing 8 kN/m³. The soil beneath the bin is homogeneous for a great depth and has a modulus of elasticity of 6000 kPa corrected for foundation shape. Determine the settlement beneath the coal at the centre of the pile and also at the edge of the stockpile along its long axis. Neglect, for this example, the effect of bin weight, foundation embedment and secondary compression.

Use the method of superposition taking advantage of the elastic solutions for a flexible foundation given in Table 8.1. The settlement, S_A, at the centre of the flexible loaded area for L/B = 25 m/5 m = 5 is calculated as follows:

$$S_A = (8 \text{ kN} \,/\, m^3 \cdot 4 \text{ m})\,(5 \text{ m})\frac{(1 - 0.35^2)\,(2.10)}{6000 \text{ kN} \,/\, m^3} = 49.1 \text{ mm}$$

Now use superposition to calculate the settlement along the long axis of the foundation at the edge of the short side. Divide the foundation into two equal size parts each 2.5 m by 25 m in size (L/B = 25 m/2.5 m = 10) (see Fig. 8.46).

Calculate the settlement at the corner of one part (i.e. at point B), and multiply that settlement by 2 to consider the equal settlement caused by both parts.

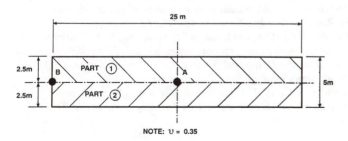

NOTE: $\upsilon = 0.35$

Figure 8.46. Explanation for example 6.

$$S_B = (2)\,(8\,\text{kN}\,/\,\text{m}^3 \cdot 4\,\text{m})\,(2.5\,\text{m})\,\frac{(1 - 0.35^2)\,(1.28)}{6000\,\text{kN}\,/\,\text{m}^3} = 30.0\,\text{mm}$$

The influence factors I_f given in Table 8.1 for the edge of the flexible, rectangular areas are for the midpoint of the long side of the foundation (refer to Note 5, Table 8.1). Therefore, the 'edge' influence factors given in Table 8.1 cannot be used to calculate the settlement of point B in this example since point B is at the midpoint of the small end of the rectangle. Influence factors for desired values of L/B not tabulated in Table 8.1 can be readily obtained by plotting the influence values given in the table for a given condition as a function of L/B.

REFERENCES

Aitchison, G.D. & Richards, B.G. 1965. A broad-scale study of moisture conditions in pavement subgrades throughout Australia and the selection of design values for soil suction equilibria and soil suction changes in pavement subgrades. In *Moisture Equilibria and Moisture Changes in Soils Beneath covered Areas*. Butterworth, Australia, pp 226-232.

Baguelin, F., Jezequel, J.F. & Shields, D.H. 1978. *The Pressuremeter and Foundation Engineering*. Trans. Tech Publications, First Edition.

Barksdale, R.D., Bachus, R.C. & Calnan, M.B. 1982. Settlements of a tower on residual soil. Engineering and Construction in Tropical and Residual Soils. *ASCE Geotech. Div. Spec. Conf. Honolulu, Hawaii*: 647-664.

Blight, G.E. 1965a. A study of effective stresses for volume change, In *Moisture Equilibria and Moisture Changes in Soils Beneath Covered Areas*. Butterworth, Australia, pp 259-269.

Blight, G.E. 1965b.. The time-rate of heave of structures on expansive clays. In *Moisture Equilibria and Moisture Changes in Soils. Beneath Covered Areas*. Butterworth, Australia, pp 78-87

Blight, G.E. 1974. Indirect determination of in situ stress ratios in particulate materials. *Proc. Engineering Foundation Conf. on Subsurface Exploration for Underground Excavation and Heavy Construction*. Henniker, N.H., USA, pp. 350-365.

Blight, G.E. 1984. Power station foundations in deep expansive soil. *Proc. Int. Conf. On Case Histories in Geotech Eng, St Louis* 1: 353-362.

Blight, G.E. 1987. Lowering of the groundwater table by deep-rooted vegetation-The geotechnical effects of water table recovery. *Proc. 9th European Conf. On Soil Mech. & Found. Eng., Dublin* 1: 285-288

Blight, G.E., Schwartz, K, Weber, H. & Wiid, B.L. 1991. Preheaving of expansive soils by flooding-failures and successes. *7th Int. Conf. On Expansive Soils, Dallas, Texas* 1: 131-135.

Brand, E.W. & Phillipson, H.B. 1985. *Sampling and testing of residual soils*. Southeast Asian Geotechnical Society, Scorpion Press, Hong Kong.

Burland, J.B., Broms, B.B. & de Mello, V.F.B. 1977. Behaviour of foundations and structures. State-of-the-Art Report. *9th Int. Conf. Soil Mech. & Found. Eng., Tokyo, Japan* 2: 495-546.

Bycroft, G.N. 1956. Forced vibrations of a rigid circular plate on a semi-infinite elastic space on an elastic stratum. *Phil. Trans. Roy. Soc., London, Series A* 248: 327-368.

Finn, P.S., Nisbet, R.M. & Hawkins, P.G. 1984. Guidance on Pressuremeter, Flat Dilatometer and Cone Penetration Testing, Site Investigation Practice: Assessing BS 5930, 20th Regional Meeting of Engineering Group of the Geological Society, University of Surrey, Guildford, 1984. *Geological Society Engineering Geology Special Publication 1*: 223-233.

Hall, B.E., Legg, P.A. & Partridge, T.C. 1991. Characteristics of a deeply weathered residual norite. In G.E. Blight et al. (eds), *Geotechnics in the African Environment* 1: 41-48. Rotterdam: Balkema.

Jaros, M.B. 1978. The settlement of two multi-storey buildings on residual Ventersdorp lava. *S.A. Instn. Civ. Engs. Symp. on Soil-Structure Interaction, Durban, South Africa,* 20pp.

Jennings, J.E. & Knight, K. 1975. A guide to construction on or with materials exhibiting additional settlement due to collapse of grain structure. *Proc., 6th Reg. Conf. For Africa on Soil Mech and Found. Eng., Durban* 1: 99-105.

Jones, D.L. & van Alphen, G.H. 1980. Collapsing sands – a case history. *Proc. 7th Reg. Conf. For Africa on Soil Mech. And Found Eng., Accra, Ghana* 2: 769-774.

Jones, G.A. & Rust, E. 1989. Foundations on residual soil using pressuremeter moduli. *Proc.12th Int. Conf on Soil Mech. And Found. Eng., Rio de Janeiro* 1: 519-524

Knight, K. 1961. The collapse of structure of sandy sub-soils on wetting, PhD Thesis, University of the Witwatersrand, Johannesburg.

Knight, K. & Dehlen, G.L. 1963. The failure of a road constructed on a collapsing soil. *Proc. 3rd Reg. Conf. for Africa on Soil Mech. And Found Eng., Salisbury* 1: 31-34.

Mair, R.J. & Wood, D.M. 1987. Pressuremeter Testing – Methods and Interpretation, CIRIA Ground Engineering Report: In-Situ Testing. Butterworths, London.

Martin, R.E. 1977. Estimating Foundation Settlements in Residual Soils. *Journal of the Geotechnical Engineering Division, ASCE* 103(GT3): 197-212. March, New York, NY.

Martin, R.E. 1987. Settlement of residual soils. *Geotechnical Special Publication* № 9, Foundation and Excavations in Decomposed Rock of The Piedmont Province, ASCE, New York, NY.

Menard, L. 1965. *The interpretation of pressuremeter test results.* Sols-Soils, Paris, 26.

Moore, P.J. & Chandler, K.R. 1980. Foundation evaluation for a high rise building in Melbourne. *Proc. 5th Southeast Asian Conf. Soil Eng., Taipei, Taiwan* 1: 245-259

Partridge, T.C. 1989. The significance of origin for the identification of engineering problems in transported Quaternary soils. In De Mulder & Hageman (eds), *Applied Quaternary Research*: 119-128. Rotterdam: Balkema.

Pavlakis, M. 1983. Prediction of foundation behaviour in residual soils from pressuremeter tests. PhD Thesis, University of the Witwatersrand, Johannesburg, South Africa.

Pellissier, J.P. 1991. Piles in deep residual soils. *Proc. 10th Reg. Conf. For Africa on Soil Mech. & Found. Eng., Maseru, Lesotho* 1: 31-40.

Pidgeon, J.T. 1980. The rational design of raft foundations for houses on heaving soil. *Proc. 7th Reg. Conf. For Africa on Soil Mech. and Found. Eng., Accra, Ghana* 1: 291-300.

Russam, K. & Coleman, J.D. 1961. The effect of climatic factors on subgrade moisture conditions, *Géotechnique* XI(1): 22-28.

Schmertmann, J.H. 1970. Static cone to compute static settlement over sand. *Journal of the Soil Mechanics and Foundation Division, ASCE* 96(SM3): 1011-1043.

Schreiner, H.D. 1987. *State of the art review on expansive soils.* Transport and Road Research Laboratory, UK.

Schwartz, K. & Yates, J.R.C. 1980. Engineering properties of aeolian Kalahari sands. *Proc. 7th Reg. Conf. For Africa on Soil Mech. & Found. Eng., Accra, Ghana* 1: 67-74.

Sellgren, E. 1981. Friction piles in non-cohesive soils. Evaluation from pressuremeter tests. PhD Thesis, Chalmer's University of Technology, Goteburg, Sweden.

Selvadurai, P., Bauer, G. & Nicholas, T. 1980. Screw plate testing of a soft clay. *Canadian Geotechnical Journal* 17(4): 465-472.

Smith, D.A. 1987a. Geotechnical application of screw plate tests, Perth, Western Australia, Vol.2. *VIII Panamerican Conference on Soil Mechanics and Foundation Engineering Cartagena*: 153-164.

Smith, D.A. 1987b. Screw plate testing of very soft alluvial sediments, Perth, Western Australia, Vol. 2. *VIII Pan American Conference on Soil Mechanics and Foundation Engineering, Cartagena*: 165-176.

Tromp, B.E. 1985. Design of stiffened raft foundations for houses on collapsing sands. Internal report, Schwartz, Tromp and Associates, Johannesburg.

Wagener, F.v.M. 1985. Personal communication with author.

Watt, I.B. & Brink, A.B.A. 1985. Movement of benchmarks at the Pienaars river survey base. In A.B.A. Brink (ed.), *Engineering Geology of Southern Africa*. Building Publications, Pretoria, South Africa, pp. 199-204.

Weltman, A.J. & Head, J.M. 1983. Site Investigation Manual. CIRIA Special Publication 25/PSA Civil Engineering Technical Guide 35.

Weston, D.J. 1980. Compaction of collapsing sand roadbeds. *Proc. 7th Reg. Conf. for Africa on Soil Mech. and Found. Eng. Accra, Ghana* 1: 341-354.

Williams, A.A.B. 1975. The settlement of three embankments on ancient residual soils. *Proc. 6th Regional Conf. for Africa on Soil Mech. and Found. Eng., Durban, South Africa* 1: 255-262.

Williams, A.A.B. 1980. Severe heaving of a block of flats near Kimberley. *Proc. 7th Reg. Conf. for Africa on Soil Mech. and Found. Eng., Ghana* 1: 301-309.

Williams, A.A.B. 1991. The extraordinary phenomenon of chemical heaving and its effect on buildings and roads. *Proc 10th Reg. Conf. for Africa on Soil Mech. & Found. Eng., Maseru, Lesotho* 1: 91-98.

Willmer, J.L., Futrell, G.E. & Langfelder, J. 1982. Settlement predictions in Piedmont residual soil. Engineering and Construction in Tropical and Residual Soils. *ASCE Geotech. Div. Spec. Conf. Honolulu, Hawaii*: 629-646.

Zeevaert, L. 1980. Deep foundation design problems related to ground surface subsidence. *Proc. 6th South East Asia Conf. on Soil Eng. Taipei, ROC* 2: 71-110.

CHAPTER 9

Shear strength behaviour and the measurement of shear strength in residual soils

R.P. BRENNER
EWI Engineers and Consultants, Zurich, Switzerland

V.K. GARGA
Civil Engineering Department, University of Ottawa, Ont., Canada

G.E. BLIGHT
Civil Engineering Department, Witwatersrand University, Johannesburg, South Africa

9.1 BEHAVIOUR AND DIFFERENCES OF RESIDUAL SOILS FROM TRANSPORTED SOILS

The selection of appropriate strength parameters and the prediction of deformation and failure are important steps in the design of structures on residual soils. In countries with residual soils, strength testing practice, both in the laboratory and in the field has mostly followed standard procedure employing triaxial and shear box tests in the laboratory and some form of penetration test, or plate load test in the field (Brand & Phillipson 1985). The need for a clear understanding of the difference between transported and residual soils arises mainly with the preparation and handling of the specimens, and with the interpretation of the test results. A knowledge of the genesis of residual soils and of the factors affecting their shear strength will enable both engineers engaged in design and in material testing to appreciate the peculiarities of these materials in their response to deformation and shear, and will thus facilitate the selection of the most suitable design values for their work.

Residual soils develop a particular fabric, grain structure and grading in place, which makes them fundamentally different from transported soils. The latter develop their fabric as a result of their mode of deposition and their stress history after deposition. In the following section. the special features which affect the stress-strain behaviour and the strength of residual soils will be summarized and discussed.

9.1.1 *Factors affecting stress-strain and strength behaviour*

Table 9.1 lists special features encountered with residual soils which are mainly responsible for the difference in stress-strain and strength behaviour in comparison with transported soils.

Stress history. After deposition, transported soils are usually subjected to increasing effective stress due to increasing depth of burial (normal consolidation). This may change when part or all of the overburden is removed (resulting in an overconsolidated state). In the case of clays, which are deposited from suspension in water, the

Table 9.1. Comparison of residual transported soils with respect to various special features that affect strength.

Factor affecting strength	Effect on residual soil	Effect on transported soil
Stress history	Usually not important	Very important, modifies initial grain packing, causes overconsolidation effect
Grain/particle strength	Very variable, varying mineralogy and many weak grains are possible	More uniform; few weak grains because weak particles become eliminated during transport
Bonding	Important component of strength mostly due to residual bonds or cementing; causes cohesion intercept and results in a yield stress; can be destroyed by disturbance	Occurs with geologically aged deposits, produces cohesion intercept and yield stress, can be destroyed by disturbance
Relict structure and discontinuities	Develop from pre-existing structure or structural features in parent rock, include bedding, flow structures, joints, slickensides etc.	Develop from deposition cycles and from stress history, formation of slickensided surfaces possible
Anisotropy	Usually derived from relict rock fabric, e.g. bedding	Derived from deposition and stress history of soil
Void ratio/density	Depends on state reached in weathering process, independent of stress history	Depends directly on stress history

stress history after deposition fully determines the void ratio and particle arrangement.

Residual soils are formed by a weathering history and the particles evolve as a result of chemical processes (e.g. leaching, precipitation, etc.). Weathering is a weakening process and may cause some vertical and lateral unloading due to the loss of mineral matter in the altering rock. This implies a progressive modification of the in situ stresses which modifies the effect of previous stresses on the structure of the weathering material. It is therefore reasonable to consider the current structure of residual soils to be in equilibrium with and associated with their current state of stress. The effect of past stresses to which they have been subjected during their formation will be small (Vaughan 1988).

Most saprolitic or lateritic residual soils behave as if over-consolidated. Hence the A-parameter during shear is usually positive, but small. In many cases, the A-value is zero or close to zero at failure. In other words pore pressures generated by undrained shear are usually relatively unimportant and may often be ignored in analysis. Figure 9.1 shows the variation of A-parameter at failure with depth for unconsolidated undrained triaxial compression of a typical residual soil, a weathered andesite lava. Note that the maximum value of A is $+0.5$, and the mean value is less than $+0.2$.

Grain/particle strength. Weathering produces soil particles (mineral grains or agglomerations of grains) with variable degrees of weakening. The particles will, therefore, display a much wider variability in crushing strength than usually encountered with transported soils.

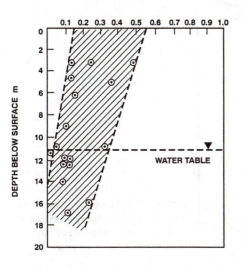

Figure 9.1. A-value profile of a residual andesite lava.

Bonding. One of the characteristics of a residual soil is the existence of bonds between particles. These bonds represent a component of strength and stiffness that is independent of effective stress and void ratio/density. Some bonding also occurs with certain transported soils (soft and stiff clays, silts, sands and even gravels) which are of great geologic age, i.e. where these bonds have had sufficient time to develop. In engineering applications these bonds are mostly ignored.

Possible causes for the development of bonds are (Vaughan 1988):
– Cementation through the deposition of carbonates, hydroxides, organic matter, etc.
– Solution and re-precipitation of cementing agents, such as silicates.
– Growth of bonds during the chemical alteration of minerals.

The strength of bonds is variable because of different minerals and differences in weathering processes. But it should be kept in mind that even a bond so weak that a sample can scarcely be handled without destroying it, provides a component of strength and stiffness which can have a significant influence on the engineering behaviour of the soil as a large mass in situ.

Figure 9.2 illustrates how the bonds in a residual soil are progressively weakened as weathering proceeds. It shows how the shear strength of a weathered andesite lava reduces progressively with decreasing bulk density as the rock weathers and progressively weakens. The bonds in the partly weathered rock are clearly of high strength (shear strength > 200 kPa) but those in the completely weathered rock are relatively weak (shear strength < 100 kPa).

Interparticle bonds are sensitive to disturbance, which is important to consider in sampling, sample preparation and shear testing. If during the initial stages of shear testing the type of test or the stress path imposed are such that significant non-uniform stresses occur prior to failure, bonds may progressively become partly or fully destroyed, resulting in a decreased measured strength. Furthermore, a bonded structure may be partly destroyed during saturation and the application of confining stresses to a test specimen if applied stresses are not carefully applied.

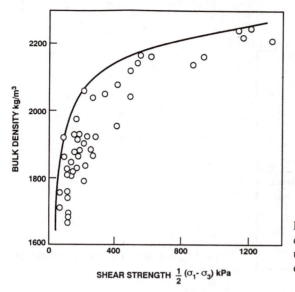

Figure 9.2. Relationship between unconsolidated-undrained shear strength under overburden stress and bulk density for residual andesite.

Remoulding of a residual soil specimen produces a 'de-structured' soil in which most bonds have been destroyed, in most cases irreversibly. Regeneration of broken bonds depends on their nature and may take from a few weeks to very long times, i.e. beyond the lifetime of engineering works (see, for example, Section 1.6).

Relict structures and discontinuities. The parent rock of a residual soil deposit may contain discontinuities of low shear strength, e.g. joints that are weathered and/or filled with low strength gouge or clay. When such seams are repeatedly and reversibly sheared (e.g. by seasonal moisture changes) they develop slickensided surfaces with a low residual strength ($\phi < 10°$). These weak zones will, after decomposition of the rock, also exist in the residual soil. They are usually difficult to discover by boring and drilling. Test specimens with such relict discontinuities will usually fail along these planes of weakness.

Many authors have concluded that the strength of saprolites may be governed almost entirely by their inherited fabric or structural features. The strength of the matrix material contained between features such as relict joints is governed by the degree of weathering and also by the degree of secondary cementation or lateritization (see Fig. 1.9). However, the overall strength of the soil mass is governed by the predominant orientation and frequency of structural features in relation to the direction of stress application, and to the strength characteristics of these features. Excellent examples of the truth of this statement have been given by Cowland & Carbrau (1988), Irfan & Woods (1988) and Lo et al. (1988).

In certain cases, it may be possible to assess the strength of the soil mass by carefully mapping the saprolitic discontinuities and measuring the strength of the discontinuity surfaces either in situ or in the laboratory. This will be possible if the discontinuities are relatively infrequent, occur in a regular pattern and are clearly defined, as in the case of a jointed rock mass. In other cases, it may be possible to characterize the relict features as a regular soil anisotropy. For example, relict bedding in a residual shale can be idealised in terms of anisotropy of strength or permeability. In such a case it is possible to take a theoretical account of the anisotropy, as shown by Florkiewicz & Mroz (1989). In many cases, however, saprolitic discontinuities are hard to characterize and their effect on soil mass strength is difficult to assess directly.

Anisotropy. As a result of stress anisotropy in a soil, the response to a shear stress application depends on the direction of the stress. In transported soils, stress anisotropy is directly associated with the mode of deposition and the stress history of the deposit. With residual soils, the anistropic behaviour has usually mainly been inherited from the fabric of the parent rock, although in situ stresses may also play a role. This applies particularly to soils derived from metamorphic rocks and where mica is present. Platy clay minerals in a decomposing rock, for example, can become oriented during the shearing process which then leads to a polished shear surface. Such surfaces may develop in situ (when not already present as relict joints), by strains accompanying soil genesis, but also by swelling and shrinkage. Due to the randomness of these processes and the residual soil fabric, it is obvious that in residual soils the stress anisotropy may vary from sample to sample.

Initial (in situ) void ratio or density. Another important property governing the engineering characteristics of residual soils is soil structure expressed by the void ratio or the density of a test specimen. The void ratio in a residual soil is a function of the stage the weathering process has reached (see Fig. 1.9) and is usually not directly related to stress history. It may vary widely and is dependent of the characteristics of the parent rock.

In a weakly bonded soil, the void ratio has a strong influence on the drained strength, which increases with density (see Fig. 9.2) (Howatt & Cater 1985, Howatt 1988). It also influences the deformation behaviour.

Partial saturation. Because of climatic conditions, groundwater tables in tropical and subtropical regions are often depressed. Evapotranspiration often potentially exceeds infiltration. This leads to deep desiccation of the soil profile. Therefore, residual soils frequently exist in an unsaturated state with continuous air in their voids. The pore air pressure will usually approximate to atmospheric pressure, but the pore water pressure will be sub-atmospheric, i.e. negative, due to capillary effects in the small pores of the soil. This negative pore water pressure or 'suction', produces an additional component of effective stress, or in other words: the effective stress becomes greater than the total stress.

The term 'suction' is expressed by:

$$(u_a - u_w)$$

where u_a = pore air pressure, u_w = pore water pressure.

The equation for the shear strength of a partially saturated soil can be written in terms of the two stress state variables $(\sigma - u_a)$, and $(u_a - u_w)$, where σ is the total stress, either as

$$\tau = c' + (u_a - u_w) \tan \phi^b + (\sigma_n - u_a) \tan \phi' \tag{9.1a}$$

(Fredlund et al. 1978), where τ = shear strength, c' = effective cohesion intercept, ϕ^b = angle of cohesion intercept increase with increasing suction, σ_n = total normal stress, ϕ' = effective angle of shear resistance, or

$$\tau = c' + [(\sigma_n - u_a) + \chi(u_a - u_w)] \tan \phi' \tag{9.1b}$$

where χ is a dimensionless modifier to the suction (Bishop & Blight 1963).

Equations (9.1a) and (9.1b) are an extension of the Mohr-Coulomb failure criterion. When a soil becomes saturated, the pore water pressure approaches the pore air pressure and Equations (9.1) take the form commonly used for saturated soils.

Figure 9.3 shows the relationship of the stress state variables $(\sigma - u_a)$ and $(u_a - u_w)$ according to Equation (9.1a) while Figure 9.4 shows the relationship according to equation (9.1b). Figure 9.4a shows a typical set of triaxial shear test results on an unsaturated soil in which u_a and u_w have been measured separately. The figure also shows the method used to determine the parameter χ. Figure 9.4b shows the same results plotted in $\frac{1}{2}(\sigma_1 + \sigma_3) - u_a$, $(u_a - u_w)$ and $\frac{1}{2}(\sigma_1 - \sigma_3)$ space. In this diagram, the slope of a line such as A'A would be the equivalent of Fredlund and Morgenstern's ϕ^b. Figure 9.4c shows typical relationships between the parameter χ in Equation (9.1b) and the degree of saturation S. The theoretical relationship was calculated for an idealized unsaturated soil consisting of an assemblage of spherical particles.

The terms ϕ^b or χ must be determined experimentally (Bishop & Blight 1962, Ho & Fredlund 1982). The value of ϕ^b is usually found to lie between 15° and 20°. χ usually falls in a range between 0 and 1, but in rare instances, may exceed 1 (see Blight 1967b). Similarly, ϕ^b may theoretically exceed 45°.

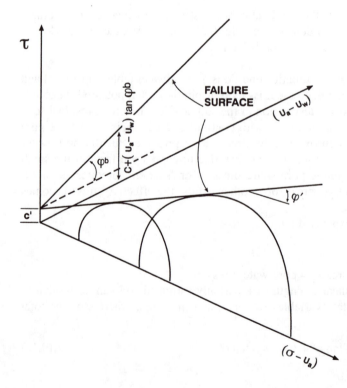

Figure 9.3. Morgenstern and Fredlund's interpretation of shear strength stress state variable relationships for an unsaturated soil.

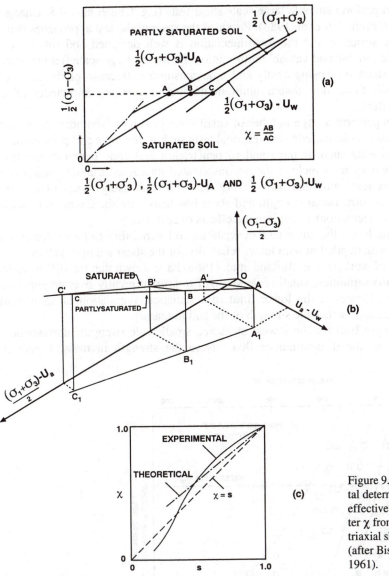

Figure 9.4. Experimental determination of effective stress parameter χ from results of triaxial shear tests (after Bishop & Blight 1961).

The evaluation of suction as a contribution to shear strength becomes important with slope stability problems in residual soils. For example, a pore water suction of, say, 100 kPa can increase the apparent cohesion of the soil by about 36 kPa.

9.1.2 *Measuring the shear strength of residual soils*

Because of the many complexities of shear strength behaviour in residual soils, two possible approaches may be adopted in measuring their strength:

1. One can perform suitable large scale shear tests (e.g. Chu et al. 1988, Chang & Goh 1988 and Premchitt et al. 1988). Provided the scale of the test approaches that of the prototype structure and the instrumentation is well designed and robust, such tests can yield reliable and valuable information. However, large scale field tests suffer the disadvantage of being costly and time-consuming. Because of the cost, it is seldom possible to do more than a minimum number of tests, and knowledge of soil variability suffers.

2. One can perform a large number of small scale in situ or laboratory tests. Suitable in situ tests could include semi-rational tests such as the vane and pressuremeter and empirical tests such as the standard penetration and cone penetrometer tests. Suitable laboratory tests include the unconsolidated undrained triaxial compression and shear box tests. Possible refinements are vane tests using vanes of different shapes to assess directional strength and shear box tests with the direction of shearing in specific orientations to explore the effects of anisotropy.

This approach has the advantage of enabling soil variability to be explored both laterally and with depth, but with lesser reliability for the shear strength values.

A number of workers, e.g. Burland et al. (1966) have shown that for stiff materials containing discontinuities, small scale strength tests may greatly over-estimate soil mass strength. However, the lower limit to a statistical population of small scale shear strengths approaches the strength of the soil in mass.

This is simply because the lowest measured small-scale strengths correspond to the strengths of the discontinuities that govern the strength in mass. Figure 9.5

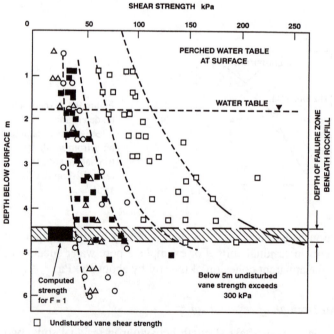

Figure 9.5. Comparison of various small scale strength measurements in a soil residual from shale with strength-in-mass calculated from failure of a waste rock dump.

□ Undisturbed vane shear strength

■ Remoulded vane shear strength

○ Unconsolidated undrained triaxial shear strength

△ Unconsolidated, quick shear box

(Blight 1969) shows a comparison of strengths measured by various small-scale methods on a lateritic residual weathered shale, with the strength-in-mass back-figured from a sliding failure in a waste rock dump founded on this material. The comparison illustrates the above statement. The scatter evident in Figure 9.5 is quite characteristic of strength measurements in residual soils. The difference between the undisturbed and the remoulded vane shear strengths is particularly marked. The undisturbed values represent the strength of material between saprolitic discontinuities. Comparing these values with the calculated strength of the soil in mass, makes it clear that the strength in mass is almost entirely governed by the strength along discontinuities. This strength is represented by the lower limit to the strength measured in small-scale tests. Various other examples of this kind will be given in what follows.

The effect of the spacing of discontinuities, joints or fissures, on measured shear strength is further illustrated by Figures 9.6a and 9.6b which both show that the measured strength for a specimen of stiff fissured soil such as London clay becomes

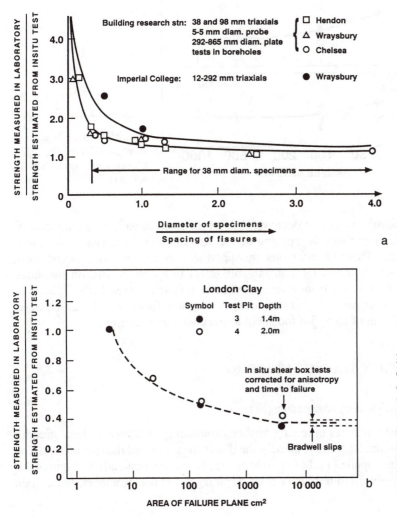

Figure 9.6. Influence of the ratio of sample size to the fissure spacing on the strength measured in laboratory tests on London clay (Marsland 1972 & Lo 1970).

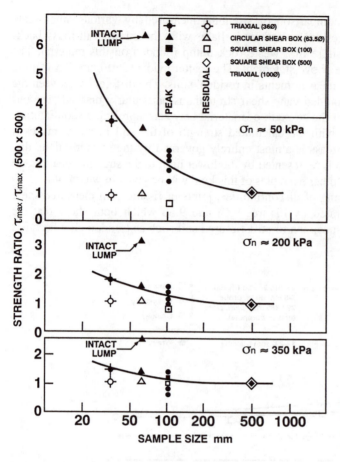

Figure 9.7. Effect of specimen size on shear strength of weathered vesicular basalt lava.

less and less realistic as its size decreases. Test specimens must have a least dimension that is 2 or more times the spacing of the discontinuities if measured strengths are to be realistic. These observations are supported, specifically for residual soils, by Figure 9.7, which shows the considerable effect of specimen size on measured strengths for a soil residual from a vesicular basalt in Brazil (Garga 1988). It is obvious from Figures 9.6 and 9.7 that the strength of a stiff fissured soil may be overestimated by a factor of up to 5 if too small a specimen size is chosen.

9.2 LABORATORY STRENGTH TESTING

9.2.1 *Types of laboratory shear strength tests*

There are two generic types of testing methods commonly used for the shear strength testing of soils in the laboratory, namely the direct shear test and the triaxial test. The systems of stresses applied in these two tests are shown schematically in Figures 9.8 and 9.9 respectively. Both tests have their advantages and disadvantages, but certain

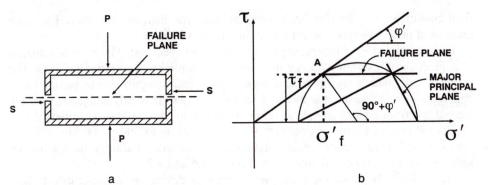

Figure 9.8. Stress system applied in direct shear , a) Forces acting on specimen, b) failure envelope through point A and Mohr's circle.

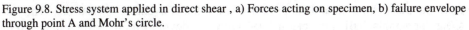

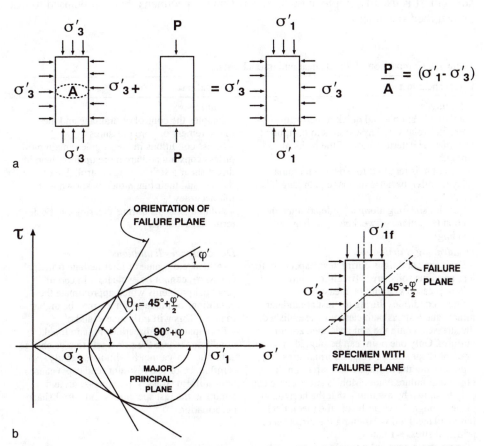

Figure 9.9. Stress system applied in triaxial compression test, a) Stresses on specimen, b) Mohr's circle and orientation of failure planes.

field conditions may be simulated better by one type than by the other. The main features of these two types of test are summarized in Table 9.2.

The triaxial test is, in theory, superior to the direct shear test. The great advantages of a direct shear test, however, are the speed and simplicity in carrying out a test. For residual soils, which are often coarse-grained or may contain coarse particles and aggregations of minerals, the direct shear test is often more advantageous than the triaxial test from a block sample as it may be easier to trim a square specimen than a cylindrical one, and with available shear box sizes of up to 300 × 300 mm, a larger specimen will give a better representation of the in situ conditions than a triaxial specimen of, say, 100 mm diameter (see Figures 9.6 and 9.7).

Figure 9.10 illustrates the principles of various devices to conduct direct shear tests, the most common and simplest one being the conventional shear box. The ring shear apparatus (Bishop et al. 1971) was developed to overcome certain disadvantages of the conventional shear box for the measurement of residual shear strength. It is usually employed with reconstituted specimens, but can be used to test undisturbed specimens.

Table 9.2. Comparison of direct shear and triaxial tests.

Direct shear test	Triaxial test
Advantages – Relatively simple and quick to perform – Enables relatively large strains to be applied and thus the determination of the residual strength – Less time is required for specimen drainage (clayey soils), because drainage path length is small – Enables shearing along a predetermined direction (e.g. plane of weakness, such as relict bedding)	*Advantages* – Enables the control of drainage and the measurement of pore pressures – Stress conditions in the sample remain more or less constant and are more uniform than in direct shear test. They are controllable during the test and their magnitude is known with fair accuracy –Volume changes during shearing can be determined accurately
Disadvantages/limitations – Drainage conditions during test, especially for less pervious soils, are difficult to control – Pore pressures cannot be measured – Stress conditions during the test are indeterminate and a stress path cannot be established, the stresses within the soil specimen are non-uniform. Only one point can be plotted in a diagram of shear stress τ versus normal stress σ, represent-ing the average shear stress on the horizontal failure plane. Mohr's stress circle can only be drawn by assuming that the horizontal plane through the shear box is the theoretical failure plane. During straining the direction of principal stresses rotates – Shear stress over failure surface is not uniform and progressive failure may develop – Saturation of fine-grained specimens (e.g. by back-pressuring) is not possible – The area of the shearing surfaces changes continuously.	*Disadvantages/limitations* – Influence of value of intermediate principal stress, σ_2, cannot be evaluated. In certain practical problems which approximate the conditions of plane strain, σ_2 may be higher than σ_3. This will influence c' and ϕ' – Principal stress directions remain fixed, conditions where the principal stresses change continuously cannot be simulated easily. – Influence of end restraint (end caps) causes non-uniform stresses, pore pressures and strains in the test specimens and barrel shape deformation

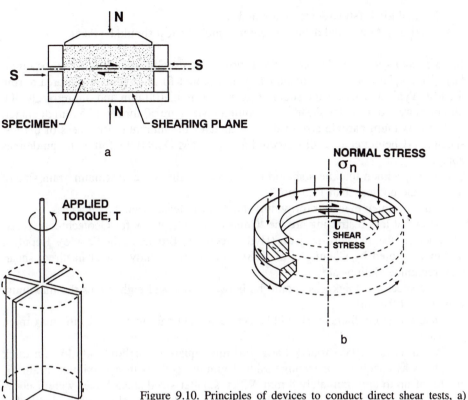

Figure 9.10. Principles of devices to conduct direct shear tests, a) Shear box, b) Ring shear apparatus, c) Vane.

In the vane shear test, direct shearing takes place between a cylindrical volume of soil and the surrounding material. Both field and laboratory vane shear devices are available. The vane can measure either the drained or undrained shear strength by suitably adjusting the rate of loading.

9.2.2 Shear box test

The results that can be obtained from a shear box test are:
- The angle of shearing resistance (peak and residual),
- The cohesion intercept (peak and residual),
- The volume change response of the soil to shearing (dilatant or contractive).

There is a large number of test variables that must be specified when planning a laboratory testing programme. However, in routine testing for engineering projects the available equipment usually limits the choice of the test variables. The following variables may have to be decided upon before starting a test series:
- Minimum size of shear box and thickness of specimen,
- Status of consolidation and drainage, saturation condition,
- Controlled strain or controlled stress test,
- Rate of shearing or stressing,

– Normal loads (stresses) to be applied,
– Maximum horizontal displacement to which the test should be taken.

Box size and specimen thickness. Shear boxes are usually square (but may be circular). With square boxes it is much easier to account for the reduction in area during the test. Typical sizes for the square box are: 60 mm, 100 mm and, more rarely 300 mm or more. For circular shear boxes common sizes are 50 mm and 75 mm diameter.

The maximum particle size of the soil dictates the minimum thickness of the test specimen (Cheung et al. 1988). According to ASTM D3080 the following guidelines apply:
– The specimen thickness should be at least six times the maximum grain size of the soil, and not less than 12.5 mm,
– The specimen diameter (or width) should be at least twice the thickness.

Somewhat less restricting are the 'Chinese Specifications for Geotechnical Tests' compiled for use in hydro-projects (Hydroelectricity Bureau of the People's Republic of China, quoted by Cheung et al. 1988). For testing sandy gravel in direct shear, their recommendations are:
– The specimen thickness should be between four and eight times the maximum grain size of the soil,
– The specimen diameter should be between eight and twelve times the maximum grain size.

Cheung et al. (1988) found that a 100 mm square shear box with 44 mm thick samples was adequate for testing residual granitic soils with a maximum grain diameter of up to approximately 8 mm. When smaller-sized shear boxes were used irregular stress-strain curves and higher shear strengths were obtained, probably because of excessive particle crushing and rearrangement within the confines of the box. (It should be noted here that shear boxes for 100 mm square by 20 mm thick specimens are available as standard equipment. The specimen thickness of 44 mm was achieved by removing the two porous stones and the lower platen).

As mentioned above, when residual soils contain discontinuities and fissures as a result of relict structure, their mass strength may be significantly affected. Garga (1988), for example, found that the drained strength of fissured dense soil (residual basalt) from 500 × 500 mm and 290 mm high shear box specimens was 1.5 to 3 times less than the strength from 36 mm diameter triaxial specimens in the normal stress range of 50 to 350 kPa (see Fig. 9.7). With relatively uniform specimens the size of the shear box was found to be less significant.

Status of consolidation, drainage and saturation conditions. Shear box specimens can remain unconsolidated and then be sheared undrained, or first be consolidated under an applied normal load and then sheared undrained or drained. Thus, the following test categories are possible: unconsolidated undrained (UU-), consolidated undrained (CU-) and consolidated drained (CD-) direct shear. With specimens of standard thickness, say 20 mm in the 100 mm square shear box, the drainage path is much shorter than in the triaxial test. This allows excess pore pressures to be dissipated fairly rapidly.

For coarse-grained soils, the CD test is the most appropriate way to shear a specimen. The result will give the drained strength parameters, c' and ϕ'. For fine-grained cohesive soils (clays and clayey silts) the UU or 'quick' test is often employed, but the CU and CD types are also possible. Since pore pressures cannot be measured or drainage controlled, and the specimens are usually not fully saturated, interpretations of test results may not be straightforward. From the UU and CU types of test, the strength parameters can be obtained only in terms of total stress and the shearing rate has to be as rapid as possible to maintain the undrained condition. With the CD type of test, the shearing rate may have to be extremely slow when the soil has a very low permeability. Even then there is no way to check whether shearing takes place without excess pore pressures, i.e. fully drained.

Tests for which drainage is allowed should be performed with the specimen fully immersed in water to eliminate the effects of capillary moisture stresses.

Controlled strain or controlled stress tests. The shear stress can be applied either by increasing the shear stress in increments and measuring the resulting displacement (stress-controlled test), or by moving the parts of the shear box relative to each other at a given displacement rate and measuring the resulting stress (strain-controlled test). Stress-controlled tests are not common in routine work. They are, however, convenient if tests are to be run at a very low displacement rate (the applied stress can easily be kept constant) and when the creep behaviour of the soil is of interest. They cannot determine the peak shear stress accurately and are unsuitable for residual strength measurements.

The strain-controlled test is easier to perform and allows the ultimate (residual) shear strength to be determined. Unless there are special reasons, direct shear tests are usually conducted in controlled-strain mode.

Rate of shearing. The shear rate to be applied depends on the drainage conditions under which the sample is to be sheared and thus on the permeability of the specimen. The nature of the shear box test in many cases does not allow the specimen to reach either a fully drained or fully undrained condition in a constant rate of shear test. However, in practice, it is usually possible to select a shearing rate such that the deviation from ideal conditions is not significant. Based on investigations by Gibson & Henkel (1954), Head (1982) recommends a time to failure, t_f, for drained direct shear tests of:

$$t_f = 12.7 \, t_{100} \tag{9.1}$$

where t_{100} is the time to 100% of primary consolidation. The value of t_{100} can be obtained by extrapolating the linear portion of the square root of time plot of the consolidation phase of the test. Equation (9.1) is based on attaining 95% pore pressure dissipation at the centre of the specimen. ASTM D3080 recommends:

$$t_f = 50 \, t_{50} \tag{9.2}$$

where t_{50} is the time required for the specimen to reach 50% primary consolidation. This equation gives essentially the same time to failure as Equation 9.1 (Cheung et al. 1988).

When t_f has been determined, the maximum permissible rate of shearing in a drained direct shear test can be estimated from:

$$\text{Rate of shearing} < \delta_f/t_f \qquad\qquad\qquad (9.3)$$

where δ_f is the horizontal displacement of the shear box at peak strength (failure). This value, however, is not known a priori and has to be estimated.

The appropriate rate of shearing is dictated mainly by pore pressure effects which may have a dominant influence with saturated specimens. The influence of the viscous component of the shear strength or of inertial effects in the test apparatus are in general small for tests carried out within the usual range of shear rates employed in drained tests. For granite residual soils (with D_{50} varying from about 0.2 to 2 mm) Cheung et al. (1988) found no significant differences in the strength parameters when the shear rate was varied from 0.007 to 0.6 mm per minute.

Normal loads to be applied. The normal pressures applied to the test specimens should generally straddle the maximum stress which is likely to occur in the ground for the design being investigated. At least four different values of normal stress should be used to define the strength envelope.

With cohesionless soils, the strength envelope usually passes through the origin, but with soils having a bonded structure, there will be a cohesion intercept. If this component of strength is of importance in an engineering application of the strength test result, tests have to be carried out with low normal loads on undisturbed, carefully handled specimens. In the normal stress range typical values of applied stresses are between about 200 and 800 kPa. Figure 9.11 shows the effect of soil bonds in a residual clay form Brazil (Vargas 1974) in producing an enhanced strength at low normal stresses.

Density of remoulded specimens. If tests are going to be carried out with remoulded specimens, e.g. with material that is intended for a compacted fill, the density (or void ratio) for testing should be defined. The angle of shearing resistance can then be evaluated as a function of the void ratio in the case of cohesionless soils.

Maximum shear displacement. The strain-controlled direct shear test is particularly useful when the relevant engineering problem requires a knowledge of the residual

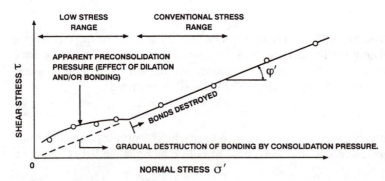

Figure 9.11. Effect of bonding on the cohesion intercept of a drained strength failure envelope (porous red clay from Sao Paolo, Brazil).

strength. The most suitable equipment to carry out such tests is the ring shear apparatus (Bishop et al. 1971), but it is not commonly available. Some commercially available shear boxes have facilities for multiple reversed shear for studying post-peak behaviour. Such devices enable horizontal displacements of any length required to reach the residual strength. The shear displacement required to reach the residual strength may be as high as 300 mm, making many reversals necessary. In order to save testing time, a faster strain rate can be used after the peak strength has been exceeded.

Tests that do not require the determination of the residual strength may be terminated after the peak strength has been passed but at not less than about 10 mm of shear displacement. With soils not showing a peak strength (which usually occurs with softer remoulded specimens) the tests should be carried to a displacement of about 15 mm.

Examples of direct shear results for residual soils: Figure 9.12 shows the results of a set of slow consolidated shear box tests on a soil residual from weathered shale at a site in KwaZulu-Natal, South Africa. The upper diagram (a) shows the development

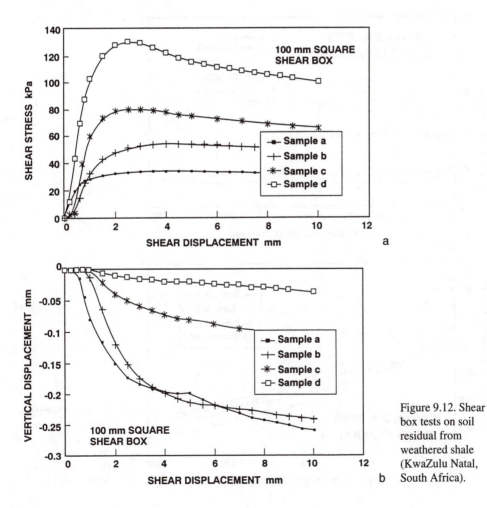

Figure 9.12. Shear box tests on soil residual from weathered shale (KwaZulu Natal, South Africa).

of shearing resistance with increasing shear displacement, while the lower diagram (b) shows the compression or contraction of the specimens as they were sheared. Figure 9.13 shows 130 mm diameter ring shear tests on the same soil. Whereas the shear box tests were on trimmed undisturbed specimens, the ring shear tests were on semi-disturbed specimens, the disturbance being caused by using a minimum amount of remoulding to form the ring specimens.

Whereas the shear box tests had to be terminated at a shear displacement of 10 mm, the ring shear tests were taken through two complete revolutions (720°), which corresponds to a shear displacement of over 400 mm. (Note that the partial disturbance in the ring shear tests removed the peaking characteristic from the shear stress – shear displacement curves). After the first stage of testing shown in Figure 9.12, the specimens were subjected to 5 reverse and forward shearing movements of 10 mm each to reach the residual shear strength (a total displacement on the shear plane of 110 mm).

The failure envelopes corresponding to Figures 9.12 and 9.13 are shown in Figure

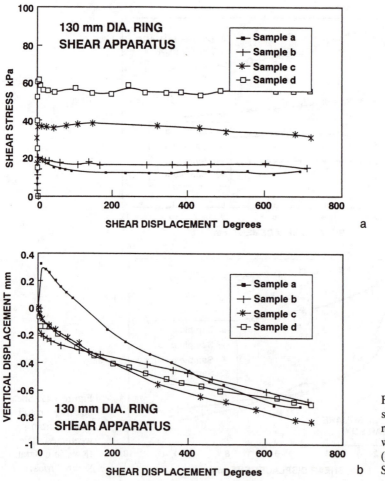

Figure 9.13. Ring shear tests on soil residual from weathered shale (KwaZulu Natal, South Africa).

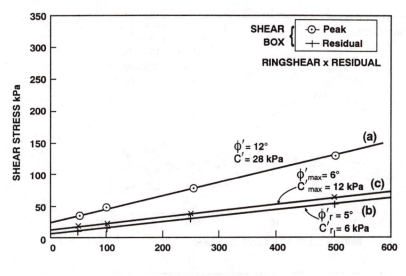

Figure 9.14. Failure envelopes for shear tests on soil residual from weathered shale (KwaZulu Natal, South Africa), a) 100mm square shear box – peak strength, b) 100mm square shear box – residual strength (5 reversals), c) 130mm diameter ring shear apparatus – residual strength.

9.14. In this case the ring shear tests gave a set of slightly larger residual shear strength parameters than did the reversed shear box tests.

Figure 9.15 shows a comparison of tests on semi-disturbed and undisturbed ring shear specimens. Here, the 'undisturbed' ring shear specimens were prepared by painstakingly cutting small blocks of undisturbed clay to fit into the ring shear apparatus. Although there were significant differences in the shear stress – shear displacement and vertical displacement – shear displacement curves, the peak and residual shear strength envelopes were very close to each other, with the 'undisturbed' specimens giving a slightly higher strength (see Figure 9.16).

9.2.3 *Triaxial test*

The triaxial testing equipment has a considerable versatility and permits of a large variety of test procedures to determine triaxial strength, stiffness and characteristic stress ratios of a soil specimen. In addition, the test can be used to measure consolidation and permeability characteristics. The state-of-the art in triaxial testing of cylindrical soil specimens was established by Bishop & Henkel (1962).

In practice, the following triaxial tests are routinely carried out, although other types of tests are also possible:

– Unconsolidated undrained (UU) test with or without pore pressure measurement,

– Isotropically consolidated undrained compression (CIU) test with or without pore pressure measurement,

– Isotropically consolidated drained compression (CID) test.

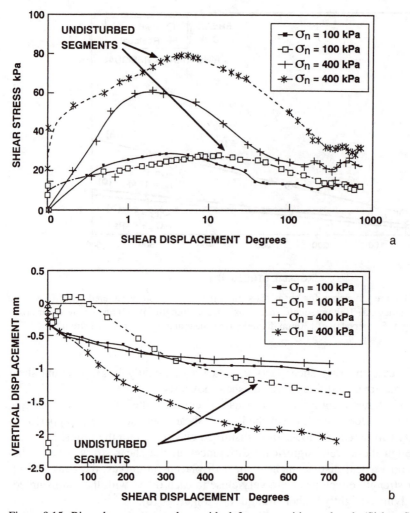

Figure 9.15. Ring shear tests on clay residual from smectitic mud-rock (Sishen South Africa). Comparison of segmental and semi-disturbed specimens.

A diagrammatic layout of the device for compression testing of cylindrical speci-mens, the triaxial cell, is shown in Figure 9.17. The specimen is sealed in a thin rub-ber membrane and subjected to a fluid pressure. In addition, a load is applied axially, through a piston acting on a top cap and controlling the magnitude of the deviator stress. Figure 9.17 shows a triaxial cell suitable for testing either saturated or unsatu-rated soils.

In a compression test the axial stress is the major principal stress σ_1. The inter-mediate and minor principal stresses, σ_2 and σ_3 respectively, are equal and corre-spond to the cell or confining pressure.

Connections to the ends of the specimen enable either the drainage of water or air from the soil voids to take place or, alternatively, the measurement of pore water pressure under conditions of no drainage. A standard test is usually carried out in two

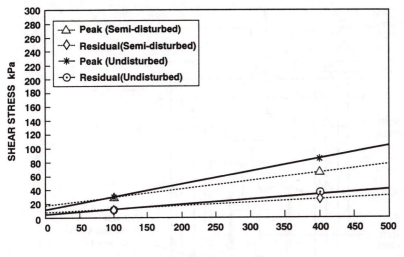

Figure 9.16. Comparison of results of ring shear tests on semi-disturbed and undisturbed specimens of clay residual from smectitic mud-rock (Sishen, South Africa).

separate stages, namely 1. The application of an all-round or confining pressure σ_3, and 2. The application of a deviator stress $\sigma_1 - \sigma_3$.

It should be pointed out that the stress systems applied to the test specimen in these tests do not necessarily agree with the stresses acting at a point in the soil for which the specimen should be representative. Also the stress paths for loading (or unloading) the specimen in the soil and in the laboratory are often not the same in routine investigation work. It follows from this that the application of triaxial test results to practical problems requires considerable interpretation. An evaluation of triaxial test results should therefore also include a knowledge of how and by what means the data have been obtained.

Triaxial test variables
The results that can be obtained from a triaxial test are, depending on the type of test and available equipment:
   – The strength envelope with peak angle of shearing resistance and cohesion intercept,
   – The pore pressure response to shearing (in undrained tests),
   – The volume change response to shearing (in drained tests),
   – Initial tangent and secant moduli (or corresponding unloading and re-loading moduli),
   – Consolidation characteristics,
   – Permeability under different confining pressures.

Sample size. For testing residual soils, the specimen diameter should not be less than 76 mm. Specimens with smaller diameters are not considered representative, because

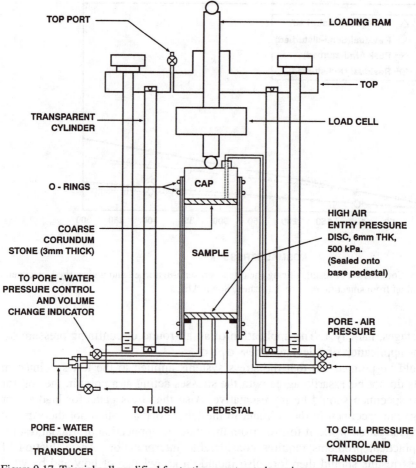

Figure 9.17. Triaxial cell modified for testing unsaturated specimens.

of the scale effect relating to fissures and joints in the soil. In addition, the specimen diameter should not be less than 8 times the maximum particle size.

The ratio of specimen length to diameter must be at least 2 to 1.

Consolidation prior to shear. The specimen is either consolidated under a specified stress system (σ_1, σ_3) prior to shear, or no consolidation is allowed (consolidated (C) and unconsolidated (U) tests, respectively). In saturated soils (clays, silts) for a series of samples taken from the same depth, the compressive strength from unconsolidated, undrained tests (or UU test) is found to be independent of the cell pressure σ_3, (with the exception of fissured clays and compact silts at low cell pressure). The strength envelope in terms of total stress is approximately horizontal, i.e. $\phi_{uu} = 0$. The undrained strength is then the apparent cohesion c_{uu}:

$$c_{uu} = \frac{1}{2}(\sigma_1 - \sigma_3)_{max} \tag{9.4}$$

which is one-half of the deviator stress at failure. For a limited range of soil types and stress paths there is a unique relationship between strength and water content at failure. The unconsolidated test is used for input in end of construction stability analyses.

For residual soils, which are often both fissured and unsaturated, the undrained strength will increase with increasing confining pressure and the strength envelope is curved. However, as the confining pressure increases, the air in the voids becomes compressed and passes into solution. Finally the stresses may be large enough to cause full saturation and ϕ_u will approach zero. Figure 9.18 shows a set of UU test

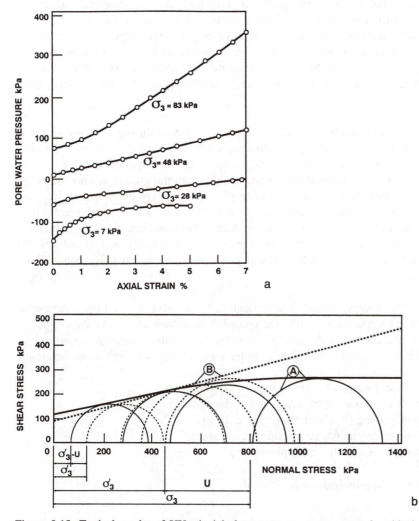

Figure 9.18. Typical results of UU triaxial shear tests on an unsaturated residual andesite soil (effective stress has been taken as $\sigma' = \sigma - u$). a) Changes in pore water pressure and effective stress during shear, b) Mohr's circles at failure drawn in terms of total and effective stresses . A – Total stress circles and failure envelopes (ϕ approaching zero as soil is saturated by compression), B – Effective stress circles and failure envelope.

results for an unsaturated residual andesite lava. Note that whereas the UU strength envelope in terms of total stresses is curved, the corresponding effective stress envelope is a straight line (Blight 1963a).

Drainage conditions. Triaxial tests are usually specified according to the conditions of drainage during the two test stages, namely:

1. Unconsolidated undrained (UU) test: No drainage and hence no pore pressure dissipation is permitted during the application of the confining pressure (all-round stress) and during the application of the deviator stress, $(\sigma_1 - \sigma_3)$ (see Fig. 9.18).

2. Consolidated undrained (CU) test: Drainage is permitted after the consolidation stress, σ_3, has been applied; the specimen fully consolidates under this stress. No drainage is permitted during the application of the deviator stress $(\sigma_1 - \sigma_3)$.

3. Consolidated drained (CD) test: Drainage is permitted throughout the test, so that full consolidation occurs under the consolidation stress σ_3 and no excess pore pressure is generated during the application of the deviator stress $(\sigma_1 - \sigma_3)$.

UU and CD tests are sometimes also designated as Q (quick) and S (slow) tests, respectively.

Consolidation stress system. The confining stress system during consolidation can either be isotropic $(\sigma_1 = \sigma_2 = \sigma_3)$ or anisotropic $(\sigma_2 = \sigma_3$ and $\sigma_1 > \sigma_3)$. In the anistropic case the additional axial stress $(\sigma_1 - \sigma_3)$ is usually applied through a load-hanger system. Since in general the stress conditions in the ground are not isotropic, i.e. $\sigma_v \neq \sigma_h$, consolidation under an anistropic stress system gives a more realistic starting point for a triaxial test than isotropic consolidation. The most frequently used anistropic stress system is σ'_1 with $\sigma'_3 = K_0\sigma'_1$, the so-called K_0 – consolidation. However, for practical applications the results of isotropically and anisotropically consolidated tests are not much different and specimens are usually consolidated isotropically.

Loading (deviator) stress system. Theoretically, it is possible to load the specimen to failure along any stress path, both in the consolidation stage and in the shearing stage (a stress path is a curve in a deviator stress versus mean stress space representing the successive stress states the specimen is subjected to during loading). Four frequently used stress paths are:
  – Compression loading (σ_1 increasing σ_3 constant)
  – Compression unloading (σ_1 constant, σ_3 decreasing)
  – Extension loading (σ_1 constant, σ_3 increasing)
  – Extension unloading (σ_1 decreasing, σ_3 constant)
Other stress paths sometimes employed are:

  – Constant mean principal stress $p' = \dfrac{\sigma'_1 + 2\sigma'_3}{3}$

  – Constant stress ratio $\dfrac{3}{2}(\sigma'_1 - \sigma'_3)/(\sigma'_1 + 2\sigma'_3) = q'/p'$

Undrained strength, effective angle of shearing resistance ϕ' and stiffness, all vary between compression and extension tests, as shown in Figure 9.19 (Baldi et al. 1988). The highest values are usually obtained with triaxial compression.

As examples of loading and unloading tests, Blight (1963c) conducted triaxial compression and extension tests on undisturbed specimens of a weakly cemented residual sand which had weathered from a micaceous schist. The strength parameters were $c' = 117$ and 93 kPa and $\phi' = 39°$ and $31°$ for the compression and extension tests respectively. These results, illustrated in Figure 9.20, show the effect of the schistose laminations in the soil. In extension, failure took place parallel to the laminations, whereas in compression the failure surfaces cut across the laminations. Note that in these tests the soil was unsaturated and u_a and u_w were measured separately (see also Fig. 9.4a).

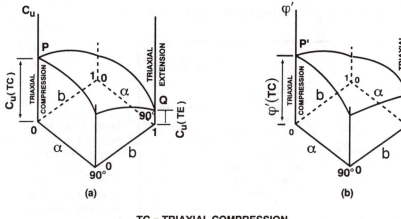

TC = TRIAXIAL COMPRESSION

TE = TRIAXIAL EXTENSION

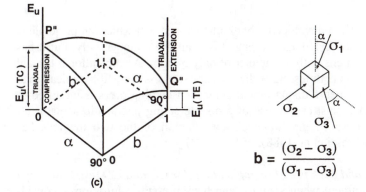

Figure 9.19. Hypothetical surfaces representing strength and stiffness parameters versus α and b axis (Baldi et al. 1988). σ_1 and σ_3 vary in single plane only and the triaxial test can furnish results only at points P and Q of the surface. a) Undrained strength, b) drained strength, c) secant stiffness at 0.1% strain.

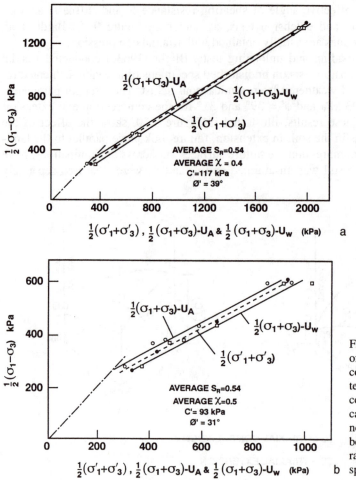

Figure 9.20. Comparison of results of triaxial compression (a) and extension (b) tests on a decomposed residual micaceous schist. Also note: The comparison between tests on saturated and unsaturated specimens.

Bishop & Wesley (1975) developed a hydraulic triaxial cell which, in principle is capable of any stress path test. This equipment is available commercially, but it has been designed for specimens with a diameter of only 38 mm. A conventional triaxial cell can be used for stress path testing if the deviator stress is applied by a dead load on the loading ram, or when a displacement-controlled loading press is modified to apply a controlled deviator stress by use of a double acting pneumatic actuator attached to the reaction bar of the frame. A combination of the two methods has proved to be practical (Baldi et al. 1988).

Saturation conditions and back pressure application (for CU and CD tests). Residual soils are usually unsaturated when sampled and back pressure saturation is considered routine in many applications.

Saturating a residual soil specimen usually represents a condition which is more severe than the natural state. In residual soil areas the ground water table is often deep and foundations well above it. In a sequence of unusually wet years it may happer that the ground water table rises by several metres. Saturation therefore represents the least favourable condition of the residual soil. It has, in general little effect on the friction angle, ϕ', but may lower the cohesion intercept and will, of course reduce the capillary suctions in the soil. Only in cases where saturation causes a weakening of existing cementation bonds in the soil will the friction angle also be reduced.

The extent to which saturation of the test specimen has been achieved can be checked by measuring Skempton's pore pressure parameter $B = \Delta u/\Delta\sigma_3$. This parameter is, however, dependent on the porosity of the specimen and on the compressibility of both soil structure and pore fluid (Black & Lee 1973). This implies that for the same B-value, the degree of saturation is not the same for soils with different stiffnesses. Higher saturation will exist in the stiffer soil.

Saturation is usually achieved by applying a back pressure. This method involves an increase in the pore pressure of the specimen thereby compressing and dissolving the gas in the specimen according to Boyle's and Henry's laws respectively. Lowe & Johnson (1973) have shown that the theoretical back pressure Δu_v required to bring a specimen from an initial degree of saturation, S_o, to a final degree of saturation, S, under conditions of constant overall soil volume and increasing water content is:

$$\Delta u_v = u_{a_0} \frac{S - S_o(1-H)}{1 - S(1-H)} \tag{9.5}$$

where u_{a_0} = initial pressure of air in voids (usually atmospheric = 100 kPa), H = Henry's coefficient of solubility (equals 0.02 volume of air per volume of water, approximately, at room temperature)

For complete final saturation ($S = 1$)

$$\Delta u_{v_{100}} = 49 u_{a_0} (1 - S_o) \tag{9.6}$$

For an initial degree of saturation of 70%, the theoretical back pressure for 100% saturation would amount to about 1450 kPa.

Single stage and multi-stage tests. In conventional triaxial testing with specimens consolidated under a selected stress system, a new specimen is set up for every test. The specimens belonging to one test series should be as identical as possible. This requirement is difficult to fulfil with most residual soils because of their great variability. In the multi-stage triaxial test, the shear strength parameters, c' and ϕ', are determined by testing a single specimen. When failure has started to develop, in the first stage the deviator stress $(\sigma_1 - \sigma_3)$ is released and the lateral stress, σ_3, is increased to a new value. Then the deviator stress is again increased until failure recommences at the new confining pressure. This procedure is repeated until sufficient failure points (usually three) have been obtained or until the deformation of the specimen is too large for reliable results to be expected (Fig. 9.21). Since only one specimen has to be set up, the multi-stage test can save testing time and thus may also be economically attractive. (However it is doubtful if small savings in the cost of testing at the expense of the quality of the test results can be regarded as justifiable).

Lumb (1964) applied the multi-stage test to drained tests on undisturbed unsaturated residual soils from Hong Kong. The results for c' and ϕ' were practically indis-

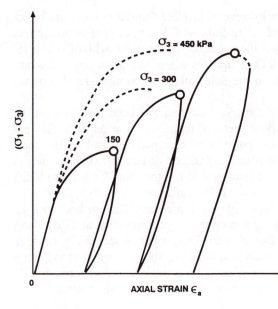

Figure 9.21. Principle of multi-stage testing: Stress-strain curve of a cyclic test.

tinguishable from the single-stage tests. He found that the maximum axial strain necessary to obtain at least three failure points was not a serious problem with undisturbed specimens. The largest total strain did not exceed 23%.

Ho & Fredlund (1982b) developed a multi-stage testing procedure for unsaturated soils to measure the increase in shear strength due to soil suction. The quantities c', ϕ' and ϕ^b can be obtained from a graphical procedure. Their procedure is set out in Figure 9.22.

Controlled strain or controlled stress tests. The usual way to shear a specimen is by applying a constant rate of axial strain. The stress-controlled test is used to simulate certain types of field loading. Lundgren et al. (1968) has discussed the advantages and disadvantages of stress-controlled shearing.

Advantages:

1. Load increments may be selected in both magnitude and duration, so that complete pore pressure equalization is obtained.

2. The deformation versus time relationship may be observed during each load increment.

3. For structurally sensitive soils an indication of the yield stress may be obtained.

Disadvantages:

1. Failure may be abrupt and result in a complete collapse of the specimen. Thus, determination of the ultimate strength or stress-strain relationship beyond the peak axial stress is not possible.

2. In drained tests, application of the failure load increment (the increment leading to failure) will usually cause failure under undrained or only partially drained conditions.

3. In undrained tests, the pore pressure induced by the failure load increment cannot be measured accurately.

STAGE	σ_3	U_a	U_w	σ_1 (peak)	σ_3-U_a	σ_1-U_a	$\dfrac{\sigma_1 - \sigma_3}{2}$	U_a -U_w	$\dfrac{\sigma_1 + \sigma_3}{2}$-$U_a$
I	241.3	103.4	68.9	725.4	137.9	622.0	242.1	34.5	380.0
II	344.8	206.9	68.9	952.3	137.9	745.4	303.8	138.0	441.7
III	448.2	310.3	68.9	1143.0	137.9	832.7	347.4	241.4	485.3

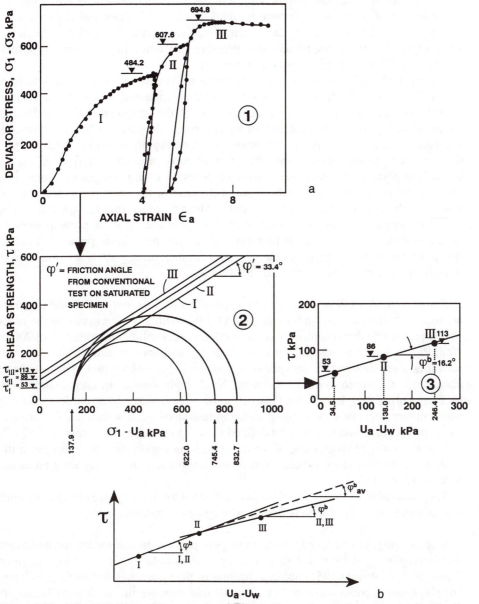

Figure 9.22. a) Determination of the strength parameters of an unsaturated specimen by Fredlund's method and multi-stage testing, b) determination of average φ^b when strength in stage III is reduced due to disturbance.

Measurement of pore water pressure during shearing. Undrained tests can be conducted with or without measuring the pore pressures. If only the value of the
undrained strength is of interest the measurement of pore pressure is not required.
Pore pressures, on the other hand, must be measured if the strength parameters in
terms of effective stress are of importance in treating the engineering problem for
which the test is conducted.

In the case of unsaturated soils with both air and water present in the voids, the
measurement of pore water pressure presents certain difficulties and requires some
additional considerations due to the effects of surface tension or capillarity. In clayey
soils, especially when compacted, this pressure difference may amount to several
hundreds of kiloPascals (Bishop & Blight 1963).

The method used to measure pore water pressure consists in bringing a rigid fine-
pored filter (normally of ceramic or sintered metal) saturated with water in contact
with the soil specimen. The pore water pressure in the specimen is then defined as
the pressure required to prevent flow of water through the porous filter. In an unsaturated soil the filter must have a sufficiently high air entry pressure to prevent air from
the specimen passing through the filter. The air entry pressure is defined as the air
pressure required to displace water from the pores of a saturated porous element. If
the air entry pressure of the porous stone at the base of the specimen is too low, air
will enter the porous stone and water will be drawn from it into the specimen. The
pressure measured on the other side of the porous element by the pressure sensing
device will then be the pore air pressure and not the pore water pressure. Recommended air entry pressures of porous ceramics for measuring pore water pressures in
triaxial tests are in the range from 200 to 600 kPa.

Cell and consolidation pressures to be applied. Cell pressures in CU and CD tests
are usually in the range from 0 to about 1200 kPa. With a back pressure in the order
of 300 to 500 kPa this results in maximum effective consolidation pressures of 700 to
900 kPa. At high stresses the angle of shearing resistance is almost insensitive to
saturation even though the soil specimen was originally unsaturated. Also effects of
disturbance are minimized. On the other hand, high stresses in combination with
back pressuring may lead to the destruction of weak bonds and/or particle crushing
which in turn decreases the angle of shearing resistance. At low stresses, the strength
envelope is curved and can only be defined by using consolidation stresses of 10 kPa
or less. Such low pressures are, of some engineering importance, for example in the
analysis of shallow slope failures, where the overburden stress may only be of the
order of 10 to 20 kPa.

The consolidation stresses applied in a test series are usually equally spaced. Four
tests or more are preferable to define the strength envelope adequately.

Rate of shearing. In undrained tests without pore pressure measurement, the undrained
shear strength, c_{uu} decreases with increasing time to failure as a result of soil creep
(Casagrande & Wilson 1951) and equalization of pore pressures within the specimen.
For clays, the decrease may be 5% for a 10-fold increase in the time to failure. A
commonly used testing rate for routine UU tests is 1 to 2% axial strain per minute.

For undrained tests (UU and CU) with pore pressure measurements, the rate of deformation must be slow enough for non-uniformities in the pore pressure distribution to equalize and in drained tests complete drainage must be allowed to occur.

In undrained tests the non-uniformity in pore pressure in the test specimen results from the non-uniformity in stress and strain due to end restraint (Blight 1963b). The pore pressure at the ends is usually higher than in the centre of the specimen. Thus, when measuring the pore pressure at the base of the specimen and using a rate that is too fast for equalization, the recorded value will be too high. This has an effect on the position of the failure envelope; the cohesion intercept being too large (Bishop et al. 1960). The time for equalization depends on the size of the specimen and the coefficient of consolidation and the drainage conditions.

Figure 9.23 shows pore pressure changes with axial strain recorded at the ends (u_e) and centre (u_c) of two triaxial compression specimens strained at a rapid rate. In both the normally and overconsolidated specimens the pore pressure recorded at the end of the specimen was greater than that recorded at the centre, and hence the effective stress would have been apparently less than the actual average value.

Figure 9.24 shows experimental equalization curves constructed from tests on two clays which were run at a range of rates of strain. The 'theoretical curves' are those calculated from the theory of Gibson & Henkel (1962). The experimental curves indicate that actual equalization of pore pressure takes place more rapidly than the theory suggests.

It is apparent from Figure 9.23 that at a particular degree of pore pressure equalization the error in the value of σ'_3 due to unequalised pore pressure will depend on the stress history of the soil. With heavily over-consolidated soils an appreciable error in the value of σ'_3 may occur even though the degree of equalization appears satisfactory. With normally consolidated soils, on the other hand, errors in σ'_3 are likely to be small, even with testing rates which are too rapid for proper equalization of pore pressure. In deciding on a test duration for an undrained test, the choice of a desirable degree of equalization must be considered in relation to the error in σ'_3 likely to occur at this degree of equalization. Lower degrees of equalization can be tolerated with soils which are normally or lightly overconsolidated. Higher degrees of equalization should be aimed at in tests on heavily overconsolidated soils (and soils which fail by shearing along a narrow zone).

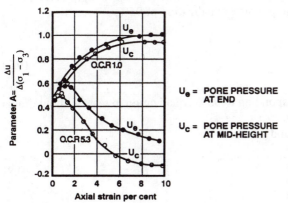

U_e = PORE PRESSURE AT END

U_c = PORE PRESSURE AT MID-HEIGHT

Figure 9.23. Typical pore pressure differences between ends and centre of triaxial shear specimens during rapid shearing (Blight 1963).

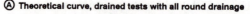

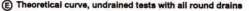

Ⓐ Theoretical curve, drained tests with all round drainage
Ⓑ Experimental curve, drained tests with all round drainage
Ⓒ Theoretical curve, drained tests with double end drainage
Ⓓ Experimental curve, drained tests with double end drainage

Ⓔ Theoretical curve, undrained tests with all round drains
Ⓖ Theoretical curve, undrained tests without drains
Ⓕ Experimental curve, undrained tests with all round drains

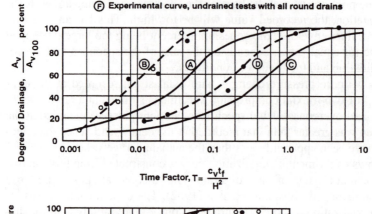

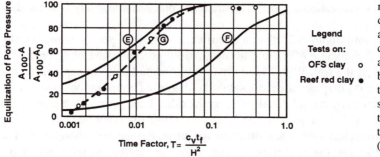

Figure 9.24. The relation between degree of drainage and time factor in drained tests (top); and the relation between equalization of pore pressure and time factor in undrained tests (bottom) (Blight 1963).

It appears, considering the pore pressure behaviour of soils with overconsolidation ratios of up to 20, that a degree of equalization of 95% is unlikely to result in an error in σ'_3 greater than 5%. An accuracy of this order is sufficient for engineering purposes. The equations below can be used to calculate the test duration, t_f, corresponding to a degree of drainage or pore pressure equalization of 95%.

For tests without drains:

$$t_f = 1.6\frac{H^2}{c_v} \tag{9.7}$$

where 1.6 is the time factor corresponding to 95% equalization in tests without drains and H is half the specimen height.

Similarly, for tests with all-round drains:

$$t_f = 0.07\frac{H^2}{c_v} \tag{9.8}$$

The meaning of the term 'test duration' will depend on the object of the test. If the object is to measure the peak shear strength parameters of a soil only, the duration of the test may be taken as the time to failure. If complete and accurate information on the whole stress path is required, the duration will be the period between the start of the test and the first significant stress and pore pressure measurement.

In Figure 9.25 Equations (9.7) and (9.8) have been presented in the form of a chart which enables a test duration giving a degree of equalization of 95% to be read off knowing c_v, the coefficient of consolidation of the soil (Blight 1963b).

Triaxial testing of stiff fissured clays. Clays formed by the in situ weathering of shales, lavas and dolerites usually contain relict joints that are commonly inclined at angles of between 45° and 60° to the horizontal. Because of the favourable orientation of the joints and fissures, failure in triaxial tests on these soils often occurs by sliding along a single inclined plane. Failure planes become visible at an axial strain of 1 to 3% and deformation usually appears to be confined to this surface with the two sections of the specimen above and below the failure surface acting as relatively rigid blocks. The axial stress on the specimen seldom falls after the appearance of the failure plane even though the contact area across the plane is continually decreasing.

Figure 9.26a shows the stress-strain curves for two typical tests on specimens of a stiff fissured clay. The specimens were enclosed by a single latex rubber membrane with a thickness of 0.15 mm and were tested in undrained compression at a strain rate of 0.05% per hour. The effective confining stress in these tests was 105 kPa. The loading rams of the triaxial cells were guided by rotating bushes so that the effects of ram friction could be assessed.

In Figure 9.26b the axial stress at which the shear planes first became visible has been allocated a relative value of 100%. The axial stresses increased considerably after the formation of the failure planes whether the ram bushes were rotating (R) or stationary (S).

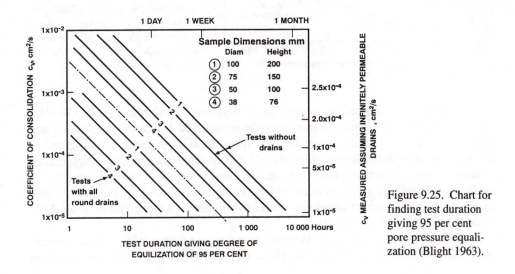

Figure 9.25. Chart for finding test duration giving 95 per cent pore pressure equalization (Blight 1963).

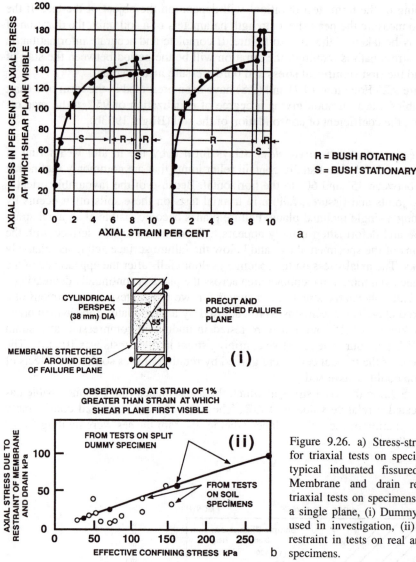

R = BUSH ROTATING
S = BUSH STATIONARY

Figure 9.26. a) Stress-strain curves for triaxial tests on specimens of a typical indurated fissured clay, b) Membrane and drain restraint in triaxial tests on specimens failing on a single plane, (i) Dummy specimen used in investigation, (ii) Measured restraint in tests on real and dummy specimens.

It appears to be generally accepted (e.g. Andresen & Simons (1960), Warlam (1960), Bishop et al. (1965) that the use of a rotating bush virtually eliminates ram friction. If this is the case, it will be seen from the difference between the stress-strain curves with the bushes stationary and rotating that even at relatively large axial strains of 5 to 10% ram friction accounted for only 5 to 20% of the measured deviator stress. The remaining increase in deviator stress after the shear planes had formed can be ascribed to restraint developed by tension in the membrane pulling against the ends of the sliding blocks of soil.

Little information is available on membrane restraint in triaxial tests where failure occurs on a single plane. Bishop & Henkel (1962) considered that the correction would not exceed 14 kPa at axial strains of 4 to 5%. However, a rough analysis of the mechanics of the problem shows that, if the membrane does not slip over the surface of the specimen, local strains in the rubber may be as much as 30 times the nominal axial strain of the soil specimen. This together with the evidence of Figure 9.26 indicates that membrane corrections for stiff fissured clay specimens might far exceed 14 kPa.

An investigation into membrane restraint was carried out with a 38 mm diameter by 76 mm high rigid dummy specimen. The specimen (see sketch in Fig. 9.26b) was cut along a diagonal plane inclined at 55° to the ends of the cylinder. The two faces of the pre-cut 'failure plane' were polished and the outer surface of the cylinder was roughened using emery cloth. After lubricating the failure surface with silicone grease, the dummy specimen was tested at a range of confining stresses using a cell with a rotating bush. The coefficient of friction along the lubricated failure plane was only 0.02 (Blight 1963b) and hence the measured axial stress in these tests was almost entirely due to membrane restraint.

Figure 9.26b (derived from triaxial tests on indurated clay specimens) shows an experimental relation between effective confining stress and the increase in axial stress over a strain interval of 1% from the strain at which a shear plane first became visible. This increase is thought to represent the effect of membrane and drain restraint and is compared with results from tests on the perspex dummy specimen. It appears from this comparison that the dummy specimen gave a fairly realistic representation of the behaviour of specimens of indurated fissured clay).

Selected measurements from the investigation are listed in Table 9.3.

Axial strain referred to in Table 9.3 is the strain that follows the first formation of a failure surface. To apply a correction for membrane restraint at any subsequent strain, this failure strain must be identified. In practice the simplest procedure is to watch the triaxial specimen closely and note the deviator stress at which a failure surface first appears. This stress is then taken as the failure stress for the specimen.

Examples of triaxial test results
Figures 9.27 to 9.29 are further examples of triaxial shear test results

Figure 9.27 shows typical consolidated undrained (CU) triaxial test results for a residual andesite lava. Figure 9.27a shows typical stress paths and a failure envelope for the soil with, inset, values of the pore pressure parameter A at failure. Figure 9.27b shows failure stress points for specimens of the same residual lava taken from

Table 9.3. Investigation into effects of membrane restraint on triaxial specimens failing on a single plane. Tests on a perspex dummy specimen.

		Membrane and drain restraint with specimen enclosed by	
Nominal axial strain (%)	Confining stress (kPa)	1 membrane 0.006 inch (0.15 mm) thick (kPa)	1 membrane + 1 wet filter paper drain (kPa)
1	70	18	27
1	280	53	91
5	70	35	56
5	280	102	130

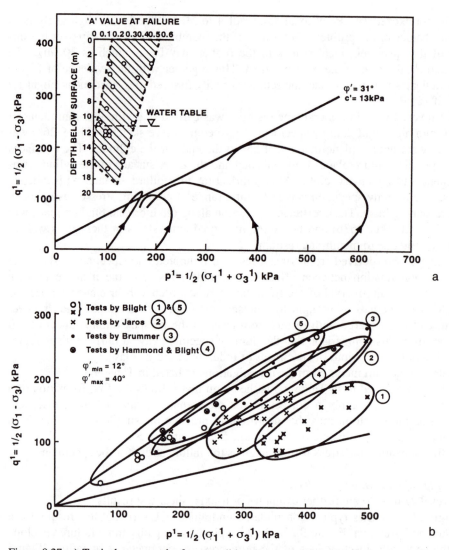

Figure 9.27. a) Typical stress paths for consolidated undrained triaxial shear of residual andesite lava, b) Summary of triaxial shear test results on undisturbed weathered andesite lava from five different sites in the Johannesburg graben. Tests were unconsolidated undrained (with measured pore pressures) on 75mm dia specimens.

5 different sites within a distance of about 10 km of each other. This illustrates the inherent variability of soil, all of which has the same origin, but which arises from a number of different lava flows and in which the degree of weathering and age also vary (Blight 1996).

Figure 9.28 (Bishop & Blight 1962) shows a set of pore pressure measurements made on consolidated undrained triaxial compression tests on specimens of unsaturated soil at increasing values of $(\sigma_3 - u_a)$. The pore air (u_a) and water $(u_w$ pressures have been measured separately in a triaxial cell equipped like the one illustrated in

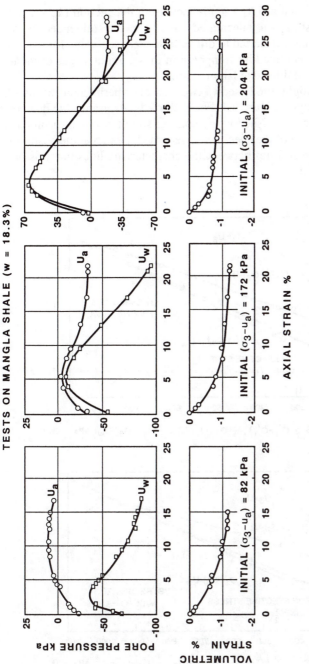

Figure 9.28. Pore pressure changes during undrained shear of a partly saturated compacted soil residual from weathered shale.

Figure 9.17. It will be noted that as $(\sigma_3 - u_a)$ is increased, the suction $(u_a - u_w)$ decreases. Also that in each test $(u_a - u_w)$ decreased initially, as the soil tended to compress and then increased as the soil started to dilate. In the test at a value for $(\sigma_3 - u_a)$ of 204 kPa, the soil became saturated by compression in the initial stages of the test and then desaturated as the soil dilated at larger strains.

Figure 9.29 (Bishop & Blight 1962) shows two sets of experimental results like those shown diagrammatically in Figure 9.4. For each of the diagrams, the degree of equalization of pore pressure established by comparing simultaneous measurements of u_a at the base and mid-height of the specimen is recorded. Note that the position of the strength line for saturated soil lies between the characteristic lines in terms of u_a and u_w for the unsaturated soil.

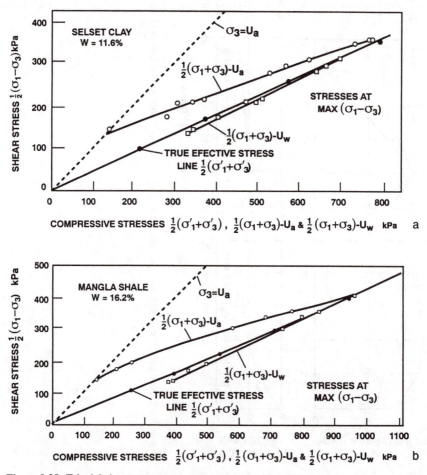

Figure 9.29. Triaxial shear at constant water content with various values of $(\sigma_3 - u_a)$ on two partly saturated soils. Degree of pore pressure equilization, a) Selset clay: 95%, b) Mangla shale: 90%.

9.2.4 *Determination of K_0 from the triaxial test*

K_0 , the coefficient of earth pressure at rest is defined by

$$K_0 = \frac{\sigma_h'}{\sigma_v'} \text{ for } \varepsilon_h = 0 \tag{9.9}$$

where the subscripts h and v refer to the vertical and horizontal directions, respectively.

The usual object of K_0 testing is to evaluate K_0 for a range of values of σ_v' and for both normally consolidated and overconsolidation conditions. Two basic ways have been used to evaluate K_0 directly in the triaxial test:

1. For zero lateral yield to occur, the change in volume of a specimen subjected to vertical compression must be equal to the horizontal cross-sectional area A multiplied by the vertical compression δ_v, i.e. if $\Delta V = A\delta_v$, $\varepsilon_h = 0$ (see inset of Fig. 9.30).

Starting with a fully consolidated specimen, the vertical stress σ_v' is increased at a rate slow enough to maintain full consolidation. δ_v and ΔV are observed and σ_h' is adjusted continuously to maintain the condition $\Delta V = A\delta_v$.

This is a very simple technique that requires no additional apparatus, and appears to give reasonably accurate results.

2. The horizontal strain ε_h may be monitored directly and kept to zero by continually adjusting σ_h' as σ_v' is varied. Various devices have been used to monitor horizontal strains. The most commonly known are the Bishop lateral strain indicator (Bishop & Henkel 1962), the strain-gauged lateral strain indicator made of a brass half-hoop, or the version of the Bishop lateral strain indicator fitted with a LVDT (linear voltage differential transformer) shown in Figure 9.30.

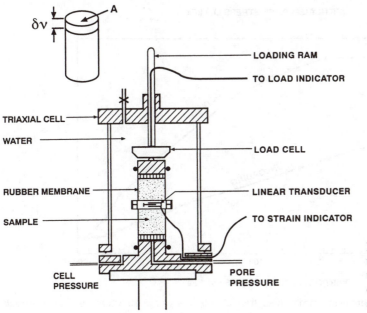

Figure 9.30. Triaxial test with a linear transducer to record or control radial strain.

Figure 9.31 shows a set of K_0 test results obtained for a soil residual from a smectitic mud-rock. (See also Figs 9.15 and 9.16). Figure 9.31a shows the measured relationship between the applied σ'_v and the corresponding σ'_h for zero lateral strain. Note that the relationship is reasonably linear for loading and that K_0 has a value of about 0.6. As the soil is unloaded, K_0 increases progressively, reaching a value of 1 when σ'_v reaches 400 kPa and 3 when $\sigma'_v = 50$ kPa.

Figure 9.31b shows the corresponding void ratio – σ'_v relationship. Note that during initial consolidation, the specimen swelled from a void ratio of 1.3 to almost 1.47 (7%). Hence the soil was heavily overconsolidated at the start of the test.

K_0 can also be measured indirectly, as indicated by Equation (8.1) and illustrated for a residual andesite lava profile by Figure 8.4.

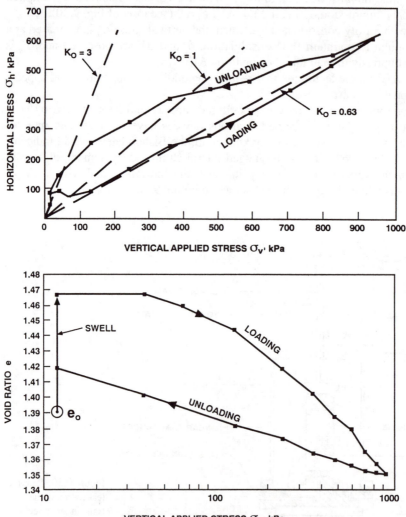

Figure 9.31. K_0 measurements made under triaxial conditions on a smectitic residual mud-rock clay.

9.3 FIELD STRENGTH TESTING

9.3.1 *General*

Field tests are advantageous for residual soils for the following reasons: 1. The disturbance caused by sampling, transportation, storage, etc. is mostly eliminated. The main factor remaining is stress release. 2. The test specimen size can be increased and becomes more representative of the soil mass. There is a variety of tests available to measure strength in situ, either directly, or by an indirect measure through empirical or semi-empirical correlations, but only a few of these have found widespread use in site investigation practice. Table 9.4 lists these tests and their limitations.

When planning a site investigation program that involves the determination of shear strength parameters, one usually has to use the locally available testing tools. These have to be used in a optimal way within the limits of the allocated budget and time, and in combination with laboratory testing. Advantages and disadvantages of the available methods must be assessed, keeping in mind which parameters are actually needed for the design and how the reliability of these parameters may actually influence the design. It is therefore essential that the geotechnical engineer have a sound understanding of the various field testing methods, their capabilities and limitations, the test variables, and of the factors which influence the test, when specifying an investigation program and later when analyzing the results.

In this section, the procedures for the field direct shear, the field vane shear, the Standard Penetration Test (SPT), and the Cone Penetrometer Test (CPT) will be described. Pressuremeter and plate bearing tests are mainly used to measure stress-deformation properties of the in situ soil. Only when they are carried to failure (which may be difficult with many residual soils) can the results be used to calculate strength parameters. These tests are therefore also discussed in Chapter 8 on settlement. The SPT and CPT are indirect tests, i.e. they do not measure the strength parameters directly.

Table 9.4. In situ strength tests.

Test	Remarks and limitations
Field direct shear test	Usually on the surface in shallow pits, time-consuming and expensive
Field vane shear test	For fine-grained soils only
Borehole shear test	Limited area of contact, multistage test, only for shallow depths
Standard penetration test	Mainly used for granular soils and stiff clays, requires pre-drilled hole
Cone penetration test	For soft/loose to medium stiff/dense, predominantly fine-grained soils
Pressuremeter (prebored, or self-bored)	For all types of soils and soft rocks; requires a high quality borehole. Difficult to use in stony soils
Plate bearing test or screw plate bearing test	Usually close to the ground surface, used mainly for settlement evaluation rather than bearing capacity, expensive

9.3.2 *Field direct shear test*

In situ direct shear tests are not frequently employed, because of their relatively high cost. Most applications reported in the literature concern rock materials, because these are usually heterogeneous and stratified and require larger test specimens to produce meaningful results. This is often true for residual soils for which discontinuities and relict joints have an important influence. The field direct shear test is particularly suitable for simulating the stress conditions that exist on a potential failure plane in a slope. It also enables shearing under the relatively low normal stress that occurs with shallow failure surfaces. Hence, field direct shear tests in residual soils are mainly employed in connection with important slope stability problems.

The main purpose of the test is to obtain the values of peak and residual strength for either the intact material or for discontinuity surfaces, including relict joints. The test is generally carried out at the bottom of shallow trenches or pits and less commonly in shafts.

Most of the tests are set up so that the shear plane will be horizontal. Ideally, the shear plane should be parallel with major discontinuity sets (e.g. relict joints) or coincide with a single major discontinuity (e.g. the shear surface of a slip).

The size of the specimen should be at least ten times the largest particle size. Typical specimen sizes are 300 × 300 mm and 500 × 500 mm for soils and weak rock.

Excavation of the test pit and of the soil pedestal (test specimen) must be done with utmost care to avoid disturbance of discontinuities in the specimen. For trimming the specimen, handsawing and cutting should be used. Once the test pedestal is exposed, it must be protected from exposure to minimize variations in water content. The final trimming must be done with a minimum of delay to avoid changes in moisture content. Special precautions are required for tests below the water table to avoid the effects of water pressure and seepage. If it is intended to shear the specimen along a specified discontinuity, the spatial orientation of this discontinuity should be carefully identified in terms of strike direction and dip before starting to trim the specimen.

The equipment for applying the normal load consists of weights, kentledge, hydraulic rams, flat jacks acting against the roof of an excavation or on an anchor system. It is important that the reaction system ensures uniform transfer of the normal load to the sample and minimum resistance to shear displacements. During the test the alignment of the normal force must be maintained, as the shear displacement increases.

The system for applying the shear force must provide a uniform load over the plane of shearing. Reactions are often provided by the excavation side walls. It is important to allow for sufficient travel in the shear force application system so that the test can be carried out without a need to reset the deformation gauges. Applied loads should be capable of being measured to an accuracy of ± 2% of the expected maximum value. Movement reference points must be well fixed and sufficiently remote from the test to ensure they are not influenced by the test forces. The effects of temperature variations on displacement should be evaluated, measured and minimized.

Examples of in situ direct shear tests

Field direct shear tests carried out on residual soils or soft rocks have been described by James (1969), Mirata (1974), Brenner et al. (1978), Brand et al. (1983) and Brand (1988), all with different shearing devices.

James (1969) reported on tests performed on sub-horizontal mudstone bands at a dam site. The test blocks were 0.6 m square. The nominal load was applied through an anchored system with the anchors taken to a depth of about 3 to 3.5 m which allowed a normal pressure of 600 to 800 kPa. Horizontal (shear) loads were applied by jacking against a concrete block built on one side of the excavation. The shearing rate was 0.5 mm/min. The test was carried out in stages. After shearing under the first normal load, the jack was released when failure became imminent. Then the next normal load was applied and time allowed for consolidation. After completion of three stages, the block was restored to its original position under a nominal vertical load. Shearing was then repeated under a nominal vertical load until the residual shear strength was reached.

Mirata (1974) introduced a special kind of shear test called the 'in situ wedge shear test'. A wedge of soil is sheared along its base by means of a single hydraulic jack (Fig. 9.32). The test has been applied in unsaturated stiff fissured clays for the

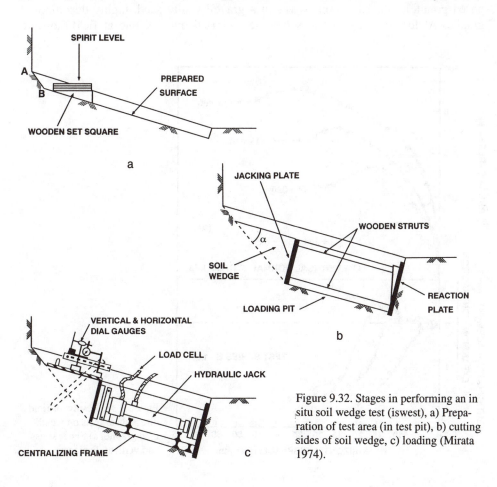

Figure 9.32. Stages in performing an in situ soil wedge test (iswest), a) Preparation of test area (in test pit), b) cutting sides of soil wedge, c) loading (Mirata 1974).

solution of slope stability problems. Its principle is to alter the inclination of the failure plane with respect to the direction of loading. In this way the ratio of shear strength to normal stress can be varied over approximately the same range of normal stress as encountered in slope stability problems. The undrained strength parameters thus obtained are used in conjunction with a total stress type of stability analysis. The test procedure is also applicable to saturated soils.

The field shear test procedure developed by Brenner et al. (1978) was designed for use in the residual soil slopes of northern Thailand. The equipment is highly portable and consists of a 305 × 305 mm and 150 mm deep shear box which derives its normal and shear reaction from a light steel frame loaded with sand bags. The shear force is applied by means of a hand-driven screw jack and proving ring assembly, which was obtained by modifying a Wykeham-Farrance field CBR apparatus. For the normal load, up to 10 kPa, a remote controlled hydraulic jack which could move freely in the horizontal plane by means of ball bearings was employed. The normal stress was measured by means of a load cell.

Typical test results obtained with this equipment are shown in Figure 9.33. These tests were carried out at two locations cut in a slope of granite residual soil with different grades of weathering. Location A was in a clayey sand (completely decomposed granite), while location B was in a gravelly, silty sand, highly decomposed granite. At location A two test series were carried out, i.e. one at field-moisture

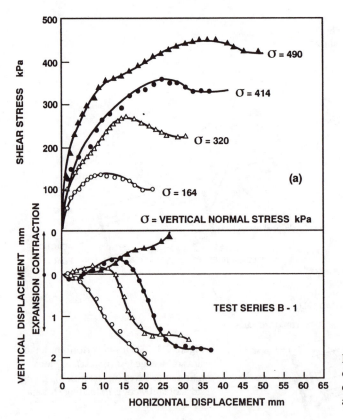

Figure 9.33. Results of field direct shear tests on weathered granite, a) Shear stress and vertical displacement.

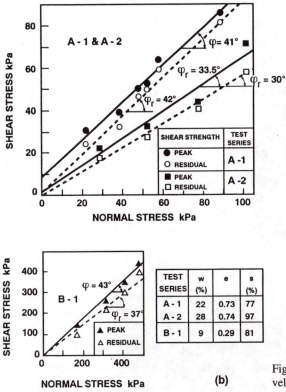

Figure 9.33. Continued. b) Strength envelopes (peak and residual).

(Series A-1) and the other soaked (Series A-2). Soaking was accomplished by lining the test pit with plastic sheeting and submerging the specimen block (which was ready to be sheared) in water for about 12 hours. At location B test series B-1 was conducted under field-moisture conditions.

Finally, the shear box described by Brand et al. (1983) and Brand (1988) was developed by the Geotechnical Control Office (GCO), Hong Kong, for use in residual soils derived from granite and volcanic rocks. The apparatus can be assembled at the test site. Each its components can be carried by one person. It has a total length of 1.75 m, a width of 0.4 m, and when assembled has a total weight of 98 kg. Figures 9.34a and 9.34b show (a) the shear box with a block of in situ soil prepared for testing, and (b) the device assembled and ready to commence a test.

Typical test results obtained with the GCO shear box are shown in Figures 9.35 and 9.36. These tests were carried out on completely decomposed granite. One series of four tests was carried out under natural water content conditions and the other under soaked conditions. It can be seen that peak shearing resistance in situ is reached at very low shear displacements (1 to 2 mm). Figure 9.15 demonstrates that the 'soaked' strength envelope is quite similar to the strength envelope obtained from triaxial tests.

a

b

Figure 9.34. a) Shear
box with specimen of
Hong Kong granite
(Brand 1988), b) In
situ direct shear test
machine developed
by Geotechnical
Control Office, Hong
Kong (Brand et al.
1983).

9.3.3 *Vane shear test*

The test is usually used only in fine grained soils which are soft enough to permit penetration and rotation of the vane blades. Hence, the typical range of application comprises low strength clayey soils, i.e. soft to medium clays which are free of stones and pedogenic inclusions. However, modified vanes have been used success-fully in soils with peak strengths up to 300 kPa (see Fig. 9.5). Sowers (1985) states that the vane test is not representative of the controlling weakness of the mass. This was well illustrated by the results shown in Figure 9.5 where the undisturbed vane shear strengths far exceeded the strength in mass of the soil, represented by the strength back-calculated from a slide failure. However, as Figure 9.5 also shows, the remoulded vane shear strength approaches the strength of a soil in mass. Possible

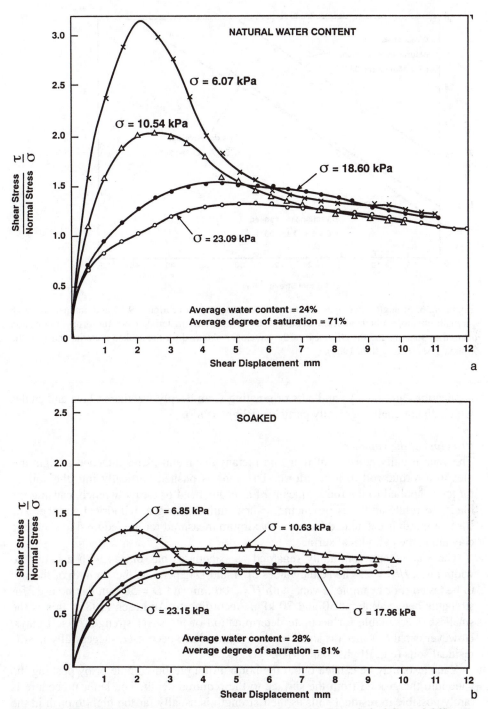

Figure 9.35. Normalized stress-displacement curves obtained with Geotechnical Control Office direct shear machine on Grade V granite, a) At natural water content, b) Under soaked conditions (Brand et al. 1983).

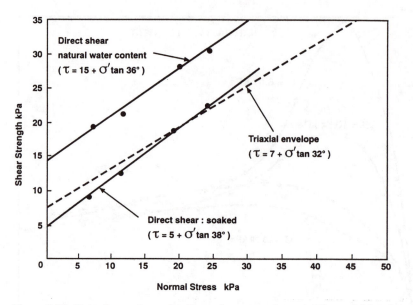

Figure 9.36. Strength envelopes for direct shear tests shown in Figure 9.14 and comparison with strength envelope obtained from triaxial tests on the same material. Note the drop in cohesion caused by soaking and the closeness of the envelopes obtained for the soaked material and with the triaxial tests (Brand et al. 1983).

applications are to clays and silts originating from deeply weathered lava and shales which do not contain gravelly particles (Blight 1985).

Principle of the vane test

The vane usually consists of four thin rectangular metal plates attached at right angles to a torque rod in a cruciform. The vane is pushed vertically into the soil. A torque is applied to the rod by means of a torque head or torque wrench, causing the blades to rotate and thus producing a shear failure along a cylindrical surface. The shear strength is calculated from the maximum measured torque required to shear the clay along the cylindrical surface.

The vane shape commonly regarded as 'standard' has blades with a height to width ratio, $H/D = 2$. The actual values of H and D depend on the strength of the soil to be tested. For example, a vane with $H = 100$ mm and $D = 50$ mm can be used for strengths between about 50 and 70 kPa. According to Andresen (1981), this is the smallest size suitable for accurate determination of the shear strength of soft clays. However, smaller vanes (as small as $H = 38$ mm) have been used successfully in stiff residual soils (e.g. Blight et al. 1970).

Vane tests may be carried out at the bottom of a pre-bored hole or by pushing the vane into the ground from the surface to the required depth. The latter procedure is rarely possible in residual soils as their strength is usually far too high to push in the vane. Figure 9.37 is a diagram of the usual vane shear arrangement for use in residual soils.

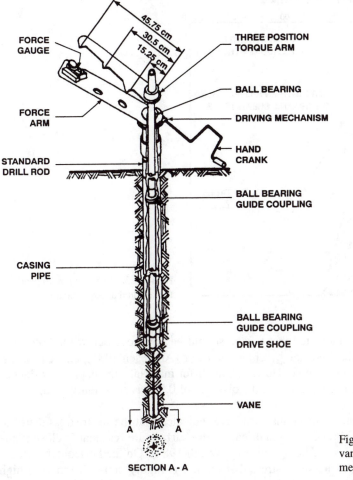

Figure 9.37. The in-situ vane shear test arrangement.

Figure 9.5 showed the results of vane shear tests in a residual shale profile, while Figure 9.38 shows a vane strength profile measured in a residual andesite lava. Note that in this profile, as with that referred to in Figure 9.5, the remoulded vane strength agrees reasonably well with the strength envelope established by means of laboratory triaxial tests, whereas the undisturbed strength is too large. This illustrates the effects of soil discontinuities, with the undisturbed vane strength representing the strength of the intact soil and the remoulded strength representing that of a fissured material.

Inherent factors of the vane test
Most experience of the use of the shear vane and research into its results has related to soft transported clays. When the vane is used in stiffer residual soils many of these effects become of negligible significance.

Effect of vane insertion. Drilling of a hole for the vane test causes disturbance by stress release below the bottom of the hole. Tests at the Norwegian Geotechnical

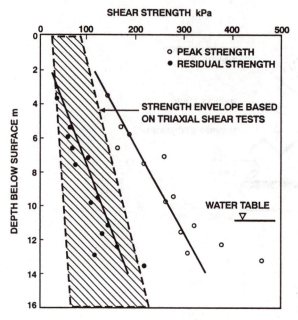

Figure 9.38. In-situ strength profile for residual andesite lava.

Institute in soft clays indicated that a vane should be advanced below the bottom of the hole to at least six times the borehole diameter (Andresen 1981), e.g. 1 m in the case of a 150 mm diameter hole. However, in stiffer materials, the depth of influence of the hole is likely to be a lot less, and a distance of 0.5 m is commonly used.

Mode of failure. Finite element analyses have indicated that the shear stress distribution on the vertical sides of the cylindrical failure surface are reasonably close to the conventional assumption of being uniform (Wroth 1984). On the horizontal surfaces of the failure cylinder, the shear stress distribution is highly non-uniform with high peaks at the edges. However, the major contribution to the measured shear strength torque arises from the cylindrical surface (86%) with the contribution by the ends being only 14%. The failure mode is most likely that of simple shear at peak shear stress, whereas the direct shear mode develops in the post-peak phase (Chandler 1988).

Shearing under undrained conditions. In order to enable a meaningful interpretation of vane test results, shearing must take place either under completely undrained conditions or completely drained conditions. Undrained conditions can be assured if, for practical purposes, the degree of consolidation $\bar{U}$ is less than 10%. The corresponding time factor has been established by Blight (1968) based on an approximate theory supported by experimental data. It takes the form:
$T = c_v\, t_f / D^2$ where c_v is the coefficient of consolidation, t_f the time to failure, and D the vane diameter. Blight's empirical equalization curve is illustrated in Figure 9.39. For a commonly used vane size of $D = 65$ mm, t_f is typically about 1 minute. From the $\bar{U}$ vs T_v theoretical drainage curve, T_v is less than 0.05 when $\bar{U} < 10\%$. Hence, c_v must be less than about 110 m²/y or 3.5×10^{-2} cm²/s for an effectively

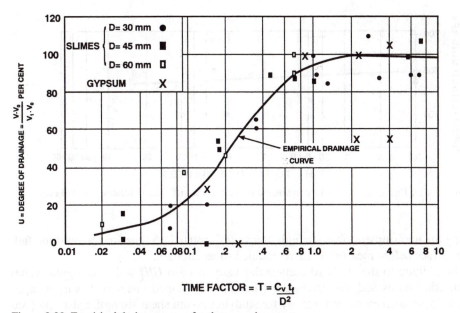

TIME FACTOR = T = $\dfrac{C_v t_f}{D^2}$

Figure 9.39. Empirical drainage curve for the vane shear test.

undrained test. This condition is usually satisfied with all clayey soils, unless the vane blade position coincides with a sand lens. Obviously, very much larger times to failure are required if a drained shear strength is to be measured. In this case, times to failure of several days are required. This type of very slow vane test has been carried out by Williams (1980) who used a slow motorized drive to rotate the vane over a period of days. In silty soils, it is possible to measure a drained shear strength within a reasonable shearing time. $\bar{U}$ will exceed 90% if $T = 0.8$. hence, if $c_v = 300$ m²/y, t_f must be at least 6 minutes.

Vane size and shape. The vane can be of any size, but usually a D size of 50 mm is used as a minimum. To minimize disturbance effects on insertion, the area ratio should not exceed 10 to 15%. (The area ratio is defined as the ratio of the cross-sectional area of the blades to the circular plan area swept by the blades).

As stated above, smaller vanes have sometimes been used in stiff clay. For example, Blight (1967a) reports on a residual fissured clay indurated by calcium and iron salts with an undisturbed vane shear strength which often exceeded 600 kPa. The vane apparatus had blades with $H = 60$ mm, $D = 30$ mm and 2 mm thick. This corresponds to an area ratio of 23%. As vane testing of very stiff soils does not appear to have been carried out previously, the effect of time to failure on vane shear strength was investigated. A series of tests was carried out at the same depth in a deposit of lime-indurated clay derived from the in situ weathering of a norite gabbro. The time to failure in these tests ranged from 5 seconds to 1 hour. The results, shown in Figure 9.40, indicated a very slight decrease in measured shear strength as the time to failure increased. This decrease, however, was less than the scatter in shear strength measurements at any one time to failure. As a result of this investigation a convenient time to failure of 1 minute has been adopted for tests on indurated clays. After meas-

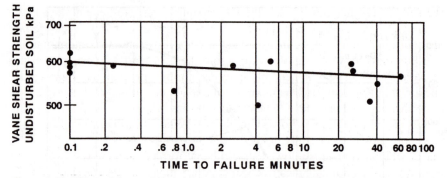

Figure 9.40. Effect of time to failure in vane shear tests on a stiff lime-indurated residual clay.

uring the undisturbed shear strength, the vane is rotated 25 times to remould the failure surface before measuring the remoulded shear strength.

In addition to the standard rectangular vane of ratio $H/D = 2$, rectangular vanes with other ratios and also triangular (rhomboidal) shaped vanes with various apex angles have been employed, mainly for studying in-situ shear strength anistropy (Aas 1965, Blight et al. 1970, Richardson et al. 1975, Silvestri & Aubertin 1988, Silvestri et al. 1993).

For calculating the shear strength ratio S_H/S_V, it is usually assumed that the shear strength is fully mobilized and uniformly distributed across the entire shear surface. The expression for the vane torque can then be written in the form:

For a rectangular vane:

$$T = \pi DH \frac{DS_v}{2} + \frac{2\pi D^2}{4} \frac{DS_H}{3} = \pi D^2 \left(\frac{HS_v}{2} + \frac{DS_H}{6} \right) \tag{9.10}$$

for a triangular vane:

$$T = S_\beta \frac{4}{3} \pi L^3 \cos^2 \beta \tag{9.11}$$

where S_v and S_H are the undrained shear strengths acting on vertical and horizontal planes respectively and S_β is the shear strength on planes inclined at angles of $\pm \beta$ to the horizontal, and L is the length of the side. Figure 9.41a shows the results of a series of vane tests using vanes of various shapes to establish the strength anisotropy in a weathered mudstone (Blight et al. 1970). In this example, S_H proved to be appreciably larger than S_v. This arose because the measurements were made in a sliding mass in which horizontal stresses parallel to the direction of sliding were considerably reduced by the slide. Figure 9.41b shows a similar set of measurements in a weathered sandstone which showed no systematic difference between S_v and S_H.

Remoulded vane shear strength: Once the undisturbed or peak shear strength has been measured, the vane is rotated 20 to 25 times and the torque is remeasured. This represents the resistance offered by the remoulded shear strength. As mentioned earlier, and illustrated in Figures 9.5 and 9.38, the remoulded strength represents the strength along an artificial fissure or joint in the soil. In stiff jointed or fissured

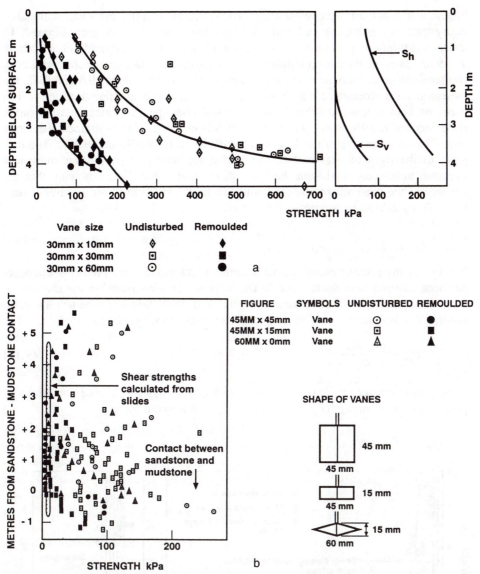

Figure 9.41. a) Directional strength of weathered mudstone at Amsterdamhoek, b) Shear strength measurements and shapes of vanes in a residual weathered sandstone.

saprolitic soils, the remoulded strength approximates to the strength of the soil in bulk (see Fig. 9.5).

Comparison of vane shear strength of residual soils with other types of measurement.

Figure 9.5 shows a comparison of vane shear strengths measured in a soil residual from weathered shale with unconsolidated undrained triaxial strengths (c_{uu}) measured on 76 mm diameter specimens and quick unconsolidated shear box tests, meas-

ured in a 76 mm diameter circular shear box. It will be seen that the small-scale laboratory tests correlate quite well with the remoulded vane shear strength. Figure 9.42
shows a similar comparison for a soil residual from a weathered mudrock (Blight et
al. 1970). Here again the remoulded vane strengths correlate reasonably well with
similar small-scale laboratory shear tests. Note that in both Figures 9.5 and 9.42, the
measured shear strengths are somewhat greater than the shear strength of the soil in
mass, as back-calculated from large-scale soil shearing movements. Also, quick
shear box test results do not differ very much from those of slow drained shear box
tests. This is indicative of the low A value of the soil. (see Section 9.1.1). The diagram to the right of Figure 9.42 shows that repeatedly sheared slow shear box
strengths agree very well with the strength calculated for slides, showing that the
strength of the mudrock in mass is controlled by the strength of the fissures it contains, i.e. by its residual strength.

9.3.4 *The pressuremeter test*

The use of the pressuremeter test to determine soil moduli and predict settlements
has been referred to in Section 8.2.4. The limit pressure measured in the test can be
used as a measure of the shear strength of the soil. In the Menard-type test, the relationship that is used to assess the unconsolidated shear strength c_{uu} is:

$$c_{uu} = \frac{p_l - \sigma_{vo}}{N_c} \qquad (9.12)$$

If the pressuremeter test is carried out sufficiently slowly for drained conditions to
prevail, a drained strength c_D can be measured, and is given by

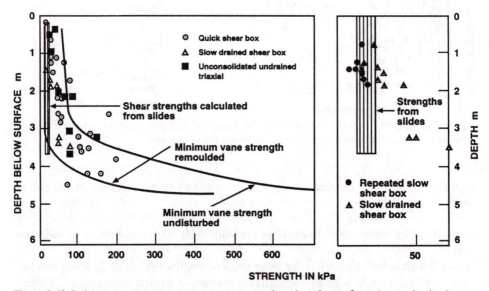

Figure 9.42. Laboratory strength measurements on weathered mudstone from Amsterdamhoek.

$$c_D = \frac{p_l - \sigma'_{vo}}{N'_c} \tag{9.12a}$$

Pavlakis (1983) has shown that if N_c is taken as 9, a good correlation between pressuremeter results and unconsolidated undrained triaxial strengths can be obtained. His results, for a very soft residual weathered siltstone, are shown in Figure 9.43. Figure 9.44 shows the pressuremeter results of Figure 8.14 interpreted as shear strengths. As many of the tests did not reach a limit pressure arrows in the diagram indicate that strengths should be higher than shown. The comparison with the triaxial shear strength envelope of Figure 9.38 shows that Equation (9.12) has some general validity.

9.3.5 *Standard penetration test (SPT)*

The appropriateness of the SPT for use in residual soils has been doubted. Blight (1985) stated that the test may at most give a rough index to soil strength or compressibility. This critique may have its justification in view that even when used for

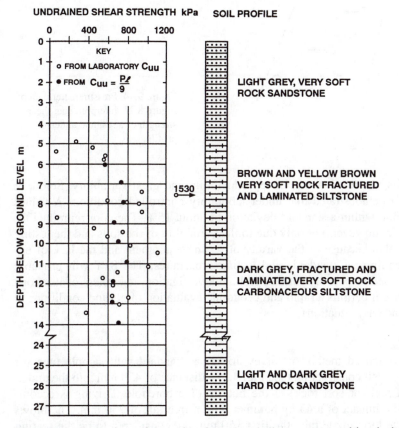

Figure 9.43. Comparison of strengths of weathered siltstone measured in triaxial and by Menard pressuremeter with $N_c = 9$.

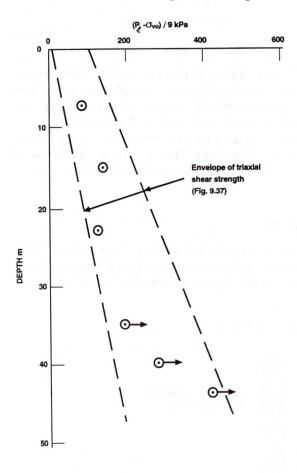

Figure 9.44. Pressuremeter data of Figure 8.14 interpreted as shear strength of a residual andesite profile.

transported soils, the test has a poor reproducibility and great variability. Serota & Lowther (1973) demonstrated that under laboratory controlled conditions *N*-values were reproducible within a standard deviation of about 15%. The poor reproducibility in field test is, however, not only due to the variability of the soils and the testing principles, but also because of the variety of testing equipment and the lack of enforcement of equipment standards and testing procedures. The SPT will probably continue to be used as part of routine borehole investigation, regardless of its shortcomings, because it permits a rapid and economic evaluation of ground conditions in both difficult and easy situations.

Principles of the test
The presently practised method involves driving a standard split sample tube (or spoon) of heavy wall construction (see Fig. 4.2) a distance of 450 mm into the undisturbed residual soil (or soft rock) at the bottom of a borehole. Driving is accomplished under the impact of a 63 kg hammer with a free fall of 750 mm. The blows required to drive the sample tube the first 150 mm are considered to be the seating drive, affected by disturbance at the bottom of the hole. The number of blows re-

quired to drive the sample tube the next 300 mm is termed the SPT '*N*' value. After driving, the sample tube is withdrawn, dismantled, and the soil sample is used for identification and index tests.

Split spoon sample tube. The thick-walled split spoon has an external diameter of 51 mm and a length of 457 mm. There is a driving shoe at one end having the same diameter as the barrel, and a coupling at the other end (see Fig. 4.2). The tube usually has a ball check valve to prevent sample loss (not shown in Fig. 4.2). Sometimes the driving shoe contains a core retainer to prevent loss of sample.

Australian, British and South African practice also has an option to fit a solid 60° cone to replace the open shoe when probing in gravelly soil in order to minimize damage to the cutting edge of the drive shoe. This was originally proposed by Palmer & Stuart (1957). The *N*-values are of similar magnitude as with the shoe, or slightly higher. When applied in loose and medium dense sands having no significant gravel content, the cone may give significantly higher *N*-values.

The SPT was originally developed to explore the properties of cohesionless transported soils. As most residual soils are cohesive, adaptations to the interpretation of the SPT have had to be made. A useful correlation has been produced by Stroud (1974) from tests on various stiff clays and soft rocks in the United Kingdom, which relates c_u/N to the plasticity index. For plasticity indices between 35 and 65% the value of c_u/N lies between 4 and 5kPa. The ratio appears to be essentially independent of depth and of discontinuity spacing. SPTs in clays may be relevant where stones prevent the extraction of undisturbed samples. Stroud (1974) also pointed out that in fissured clays the mass shear strength is only about one half of the shear strength of the intact material. The approach taken by Stroud has been found to be applicable to clayey residual soils and is quite widely used in South Africa. Stroud's correlation provides a first approximation to the mass shear strength of a residual soil and is taken as:

$$c_u = 5N \text{ kPa} \tag{9.13}$$

The SPT *N* value has also been correlated with the elastic modulus of clayey residual soils by the equation

$$E = 200\, c_u = 1000\, N \text{ kPa, or } E = N \text{ MPa} \tag{9.14}$$

Figure 9.45a shows the variation of SPT *N* with depth for a number of sites on residual andesite lava. As indicated on Figure 9.45a, the correlation with strength is reasonably good at depths of up to 5 m. At greater depths than this, Stroud's expression appears to overestimate the shear strength. However, another set of data for the same weathered andesite (Pavlakis 1985) in which the SPT tests were taken to much greater depths (50 m as opposed to 14 m) shows a much better correlation with depth (see Fig. 9.45b). This is clearly because the depth of weathering on the sites represented in Figure 9.45b was greater than for the sites represented in Figure 9.45a. This illustrates one of the many pitfalls of making generalized assumptions about residual soil profiles.

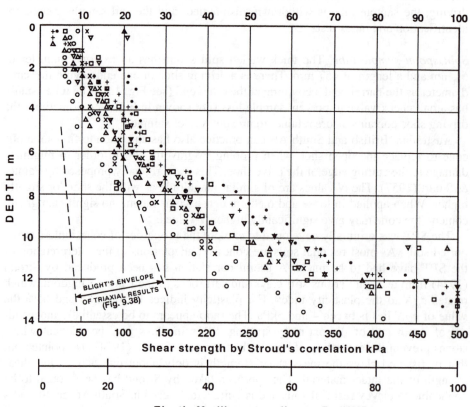

Figure 9.45. a) Variation of SPT '*N*' with depth for 8 holes in weathered residual andesite lava.

9.3.6 *Cone penetration test (CPT)*

The quasi-static cone penetration test, or CPT has been applied in residual soils to a limited extent. Residual soils are often very stiff and dense and penetration may be limited to the top few metres. Moreover, corestones and pedogenic inclusions such as lime nodules, which often occur in residual soil profiles cannot be penetrated or may deflect the cone. Still, the survey by Brand & Phillipson (1985) showed that the quasi-static CPT is fairly widely used in residual soils, mainly for shallow foundation and pile design.

Field penetrometer testing of residual soils

As the modified vane apparatus has an upper limit of shear strength measurement of 600 kPa, soils with an undisturbed vane strength exceeding this limit must be tested by an alternative means. One instrument that has been adopted for this purpose is the cone penetration test. The most widely used mechanical penetrometer tips are the Delft mantle cone and the Begeman friction cone (Fig. 9.46). Both have a mantle of slightly reduced diameter attached above the cone. A sliding mechanism allows the

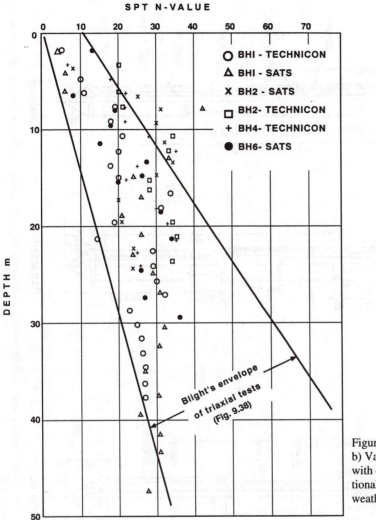

Figure 9.45. Continued. b) Variation of SPT '*N*' with depth at two additional sites of residual weathered andesite lava.

downward movement of the cone in relation to the push rods. The force on the cone (penetration resistance) is then measured as the cone is pushed downward by applying a thrust on the inner rods. If the cone is equipped with friction sleeve, a second measurement is taken when the flange engages in the friction sleeve and cone and friction sleeve are pushed down together a further increment.

In the electrical penetrometer tip the cone is fixed and the cone resistance and the force on the friction jacket are measured by means of a force transducer attached or built into the cone. Electric cables threaded through the push rods or other suitable means (e.g. solid state memory) transmit the transducer signal to a data recording system. The electrical cone penetrometer tip permits continuous recording of the quantities measured over each push rod-length interval.

Electrical piezometers are superior to the mechanical type. They permit of measurements with higher precision and with optimum repeatability. They also allow the

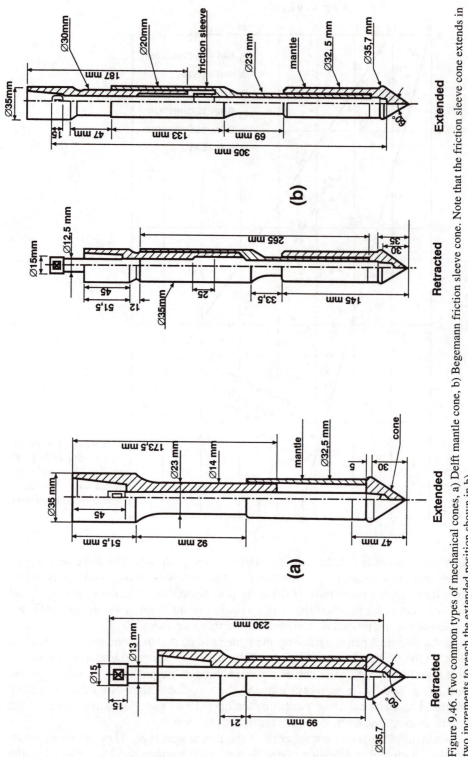

Figure 9.46. Two common types of mechanical cones, a) Delft mantle cone, b) Begemann friction sleeve cone. Note that the friction sleeve cone extends in two increments to reach the extended position shown in b).

simultaneous recording of cone resistance and friction. There is the option to install an inclinometer into the cone to check the verticality of the sounding. The string of rods may undergo considerable bending resulting in a drift of the penetrometer tip when exceeding a certain depth or when encountering an obstacle, such as a boulder or corestone.

A more recent development is the piezocone which incorporates a pore pressure transducer within the tip of an electrical cone. Pore pressure transducers for this purpose should have a volume factor of less than 2.5 mm³/kPa (De Ruiter 1982). Figure 9.47 illustrates the Fugro piezocone penetrometer. Examples of the specific use of the cone penetrometer in residual soils are given by Peuchen et al. (1996)

The relation between the undrained shear strength c_u and the cone penetration resistance q_c is of the form:

$$c_u = \frac{q_c - \sigma_{vo}}{N_c} \qquad (9.15)$$

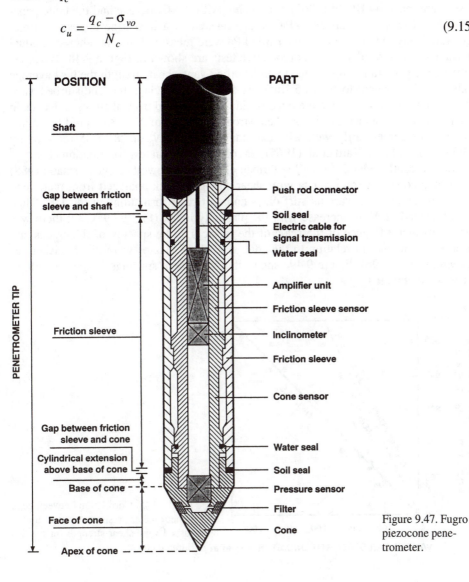

Figure 9.47. Fugro piezocone penetrometer.

where σ_{vo} is the total overburden stress at the depth of measurement, and N_c is a bearing capacity factor which must be evaluated empirically, σ_{vo} is usually negligible in comparison with q and may be omitted from Equation (9.15) with little error. Penetrometer measurements in stiff soils are usually made from the bottom of a 100 mm diameter augered hole. The probe is jacked in its closed position to a distance of 0.5 m below the bottom of the hole. The cone is then advanced separately until a steady penetration resistance is recorded. No study of the effects of the rate of penetration has so far been made, but as a steady resistance is usually reached after a penetration of 25 mm a rate of penetration of 25 mm/minute has arbitrarily been used. This gives some similarity between times to failure in the penetrometer and vane tests. The value of N_c appears to be site- and soil-specific.

To evaluate N_c for stiff residual clays in South Africa, a series of comparative tests were made by Blight (1967) at two sites with the field vane and the cone penetrometer. Strengths of up 280 kPa were measured in a lacustrine deposit of lime-indurated clay, while strengths over 300 kPa were measured in a weathered residual norite clay. The results of the comparative tests are shown in Figure 9.48. When relating cone penetration resistance to undisturbed vane shear strength the best average value for N_c appears to be 15.5. In most stiff fissured clays the undisturbed vane shear strength is about twice the unconsolidated undrained triaxial strength. Hence in relating cone resistance to triaxial shear strength a value of $N_c = 30$ would be appropriate. This agrees fairly well with extreme values for N_c of 28 to 35 reported by Thomas (1965) and Ward et al. (1965) for cone penetration tests on London clay.

More recently, Marsland & Quarterman (1982) and Powell & Quarterman (1988) correlated cone penetration data with shear strengths back-analyzed from plate load results obtained from various stiff clays and soft rock formations in the UK. They found a trend of N_c to increase with the plasticity index (Fig. 9.49a), but there was also a distinct influence of the scale of the fabric, i.e. the spacing of the cracks and fissures in the clay in relation to the cone size (three classes were distinguished, as shown in Fig. 9.49b). Figure 9.49a shows that with large spacings N_c values as high as 30 were obtained.

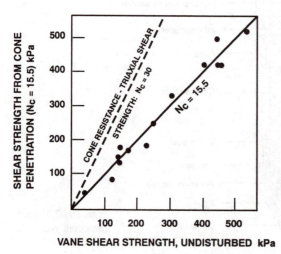

Figure 9.48. Correlation between cone penetrometer resistance and undisturbed vane shear strength for two indurated clays.

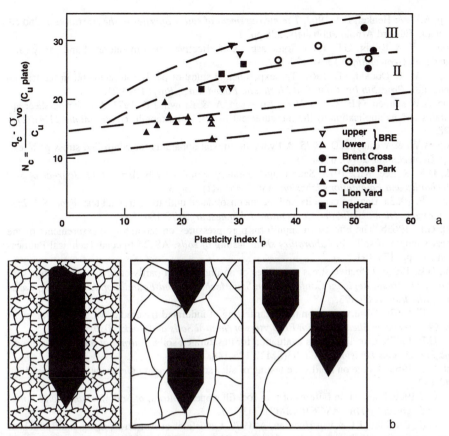

Figure 9.49. a) Cone factors for stiff clays and soft rock based on plate load tests, b) three classes of fabric features with relation to cone size (Marsland & Quaterman 1988).

REFERENCES

Aas, G. 1965. A study of the effect of vane shape and rate of strain on the measured values of in situ shear strength of clays. *Proc. 6th Int. Conf. Soil. Mech. and Found. Eng., Montreal* 1: 141-145.

Andresen, A. & Simons, N.E. 1960. Norwegian triaxial equipment and technique. *Research conference on shear strength of cohesive soils, Boulder, USA*, pp. 696-9.

Andresen, A. 1981. Exploration, sampling and in situ testing of soft clay. In E.W. Brand & R.P. Brenner (eds), *Soft Clay Engineering*. Amsterdam: Elsevier, pp. 241-308.

Baldi, G., Hight, D.W. & Thomas, G.E. 1988. Reevaluation of conventional triaxial test methods. In R.T. Donaghe, R.C. Chaney & M.L. Silver (eds), *Advanced Triaxial Testing of Soil and Rock, ASTM STP 977*. American Soc. for Testing and Materials, Philadelphia, PA, pp. 219-263.

Bishop, A.W., Webb, D.L. & Lewin, P.I. 1965. Undisturbed samples of London clay from the Ashford Common shaft: Strength-effective stress relationships. *Géotechnique*, London, England 15(1): p 6.

Bishop, A.W., Alpan, I., Blight, G.E. & Donald, I.B. 1960. Factors controlling the strength of partly saturated cohesive soils. *Proc. Research Conf. Shear Strength of Cohesive Soils, Boulder, Co., ASCE.* pp. 503-532.

Bishop, A.W. & Henkel, D.J. 1962. *The measurement of soil properties in the triaxial test*, 2nd ed. London: Edward Arnold Publishers, 228 pp.

Bishop, A.W. & Blight, G.E. 1963. Some aspects of effective stress in saturated and partly saturated soil. *Géotechnique* 13(3): 177-197.

Bishop, A.W. & Donald, I.B. 1961. The experimental study of partly saturated soil in the triaxial apparatus. *Proc. 5th Int. Conf. Soil Mech. and Found. Eng., Paris* 1: 13-21.

Bishop, A.W., Green, G.E., Garga, V.K., Anderson, A. & Brown, J.D. 1971. A new ring-shear apparatus and its application to the measurement of residual strength. *Géotechnique* 21(4): 273-328.

Bishop, A.W. & Wesley, L.D. 1975. A hydraulic triaxial apparatus for controlled stress path testing. *Géotechnique* 25(4): 657-670

Black, D.K. & Lee, K.L. 1973. Saturating laboratory samples by back pressure. *Journal of Soil Mechanics and Foundations Division, ASCE* 99(SM1): 75-93

Blight, G.E. 1963a. Bearing capacity and the unconsolidated undrained triaxial test. *Proc. S.A. Inst. Civ. Engrs., Diamond Jubilee Convention, Johannesburg*: 177-184.

Blight, G.E. 1963b. The effect of nonuniform pore pressures on laboratory measurements of the shear strength of soils. In *Laboratory shear testing of soils*. ASTM Special Technical Publication 361. pp. 173-191.

Blight, G.E. 1963c. Effective stress properties of an undisturbed partly saturated, micaceous soil. *Proc. 3rd African Regional Conference on Soil Mechanics and Foundation Engineering, Salisbury, Rhodesia* 1: 69-173.

Blight, G.E. 1967a. Observations on the shear testing of indurated fissured clays. *Proc. Geotechnical Conference on Shear Strength Properties of Natural Soils and Rocks, Oslo* 1: 97-102.

Blight, G.E. 1967b. Effective stress evaluation for unsaturated soils. *Journal of the Soil Mechanics and Foundations Division, ASCE* 93(SM2): 125-148.

Blight, G.E. 1968. A note on field vane testing of silty soils. *Canadian Geotechnical Journal* 5(3): 142-149.

Blight, G.E 1969. Foundation failures of four rockfill slopes. *Journal of Soil Mechanics and Foundation Engineering Div., ASCE* 95(SM3): 743-767.

Blight, G.E., Brackley, I.J. & van Heerden, A. 1970. Landslides at Amsterdamhoek and Bethlehem – an examination of the mechanics of stiff fissured clays. *The Civil Engineer in South Africa, June*: 129-140.

Blight, G.E. 1985. Residual soils in South Africa. In E.W. Brand & H.B. Phillipson (eds), *Sampling and Testing of Residual Soils, a Review of International Practice. Technical Committee on Sampling and Testing of Residual Soils*. Int. Soc for Soil Mechanics and Foundation Engineering, Scorpion Press, Hong Kong. pp. 159-168.

Blight, G.E. 1996. Properties of a soil residual from andesite lava. *4th Int. Conf. on Tropical Soils, Kuala Lumpur, Malaysia* 1: 575-580.

Brand, E.W., Phillipson, H.B., Borrie, G.W. & Clover, A.W. 1983. In situ shear tests on Hong Kong residual soil. *Proc. Int. Symp. Soil and Rock Investigations by In-situ Testing, Paris* 2: 13-17.

Brand, E.W. & Phillipson, H.B. 1985. *Sampling and testing of residual soils*. South East Asian Geotechnical Society, Scorpion Press, Hong Kong.

Brand, E.W. 1988. Some aspects of field measurements for slopes in residual soils. Proc. *2nd Int. Symp. Field Measurements in Geomechanics, Kobe* 1: 531-545.

Brenner, R.P. Nutalaya, P. & Bergado, D.T. 1978. Weathering effects on some engineering properties of granite residual soil in northern Thailand. *Proc. 3rd Congress Int. Assoc. Engineering Geology, Madrid*. Section 2, 1: 23-36.

Burland, J.B., Butler, F.G.B. & Dunican, P. 1966. The behaviour and design of large diameter bored piles in stiff clay. *Symp. Large Bored Piles. Instn. Civ. Engrs., London*: 51-71.

Casagrande, A. & Wilson, S.D. 1951. Effect of rate of loading on the strength of clays and shales at constant water content. *Géotechnique* 2: 251-264.

Chandler, R.J. 1988. The in-situ measurement of the undrained shear strength of clays using the field vane. In A.F. Richards et al. (eds), *Vane Shear Strength Testing of Soils: Field and Laboratory Studies*. ASTM-STP 1014, Philadelphia, PA. pp. 13-44.

Chang, M.F. & Goh, A.T.C. 1988. Laterally loaded bored piles in residual soils and weathered rocks. *Proc 2nd Int. Conf. Geomechanics in Tropical Soils, Singapore* 1: 303-310.

Cheung, C.K., Greenway, D.R. & Massey, J.B. 1988. Direct shear testing of a completely decomposed granite. *Proc. 2nd Int. Conf. Geomechanics in Tropical Soils, Singapore* 1: 109-117.

Chu, B.L., Hsu, T.W. & Lai, T.C. 1988. In situ direct shear tests for lateritic gravels in Taiwan. *Proc 2nd Int. Conf. on Geomechanics in Tropical Soils, Singapore* 1: 119-126.

Cowland J.W. & Carbray A.M. 1988. Three cut slope failures on relict discontinuities in saprolitic soils. *Proc. 2nd Int. Conf. Geomechanics in Tropical Soils Singapore* 1: 253-258.

De Ruiter, J. 1982. The static cone penetration test. State-of-the-art-report. *Proc. 2nd European Symp. on Penetration Testing (ESOPT-2), Amsterdam* 2: 389-405.

Florkiewicz, A. & Mroz, Z. 1989. Limit analysis for cracked and layered soils. Proceedings, *12th Int. Conf. on Soil Mech. and Found. Eng., Rio de Janeiro, Brazil* 1: 515-518.

Fredlund, D.G., Morgenstern, N.R. & Widger, R.A. 1978. The shear strength of unsaturated soils. *Canadian Geotechnical Journal* 15(3): 313-321.

Garga, V.K. 1988. Effect of sample size on shear strength of basaltic residual soils. *Canadian Geotechnical Journal* 25: 478-487.

Gibson, R.E. & Henkel, D.J. 1954. Influence of duration of tests at constant rate of strain on measured 'drained' strength. *Géotechnique* 4(1): 6-15.

Head, K.H. 1982. Manual of soils laboratory testing, Vol. 2: *Permeability, shear strength and compressibility tests*. Pentech Press, 747 p.

Ho, D.Y.F. & Fredlund, D.G. 1982a. Increase in strength due to suction for two Hong Kong soils. *Proc. Conf. Engineering and Construction in Tropical and Residual Soils, Honolulu, Hawaii, ASCE*: 263-295.

Ho, D.Y.F. & Fredlund, D.G. 1982b. A multistage triaxial test for unsaturated soils. *Geotechnical Testing Journal* 5(1/2): 18-25.

Howatt, M.D. 1988. The in situ strength of saturated decomposed granite. *Proc. 2nd Int. Conf. Geomechanics in Tropical Soils, Singapore* 1: 311-316.

Howatt, M.D. 1988. Written discussion in reply to the questions raised by Rocha & Queiroz: The in situ strength of saturated decomposed granite. *Proc. 2nd Int. Conf. Geomechanics in Tropical Soils, Singapore* 2: pp 603.

Howatt, M.D. & Cater, R.W. 1982. Passive strength of completely weathered granite. *Proc. 1st Int. Conf. Geomechanics in Tropical, Lateritic and Saprolitic Soils, Brasilia* 2: 371-379.

Irfan, T.Y. & Woods, N.W. 1988. The influence of relict discontinuities in saprolitic soils. *Proc. 2nd Int. Conf. Geomechanics in Tropical Soils, Singapore* 1: 267-276.

James, P.M. 1969. In situ shear test at Muda Dam. *Proc. Conf. In Situ Investigations in Soils and Rock, London*: 75-81.

Lo, K.W., Leung, C.F., Hayata, K & Lee, S.L. 1988. Stability of excavated slopes in the weathered Jurong. *Proc. 2nd Int. Conf. Geomechanics in Tropical Soils, Singapore*, 1: 277-285.

Lowe III, J. & Johnson, T.C. 1960. Use of back pressure to increase degree of saturation of triaxial test specimens. *Proc. Research Conf. Shear Strength of Cohesive Soils, Boulder, Co., ASCE*: 819-836.

Lumb, P. 1962. The properties of decomposed granite. *Géotechnique* 12: 226-243.

Lumb, P. 1964. Multi-stage triaxial tests on undisturbed soils. *Civil Engineering and Public Works Review* 59: 591-595.

Lundgren, R., Mitchell, J.K. & Wilson, J.H. 1968. Effects of loading method on triaxial test results. *J. Soil Mech. and Found. Div., ASCE* 94(SM2): 07-419.

Marsland, A. & Quarterman, R.S.T. 1982. Factors affecting the measurement and interpretation of quasi-static penetration testing in clays. *Proc. 2nd European Symp. Penetration Testing, ESOPT-II, Amsterdam* 2: 697-702.

Mirata, T. 1974. The in situ wedge shear test – a new technique in soil testing. *Geotechnique* 24: 311-332.

Palmer, D.J. & Stuart, J.G. 1957. Some observation on the standard penetration test and the corre-lation of the test in situ with a new penetrometer. *Proc. 4th Int. Conf. Soil. Mech and Found. Eng., London* 1: 231-236.

Pavlakis, M. 1983. Prediction of foundation behaviour in residual soils from pressuremeter tests. PhD Thesis, Witwatersrand University, Johannesburg.

Peuchen, J., Plasman, S.J. & van Steveninck, R. 1996. In situ testing of tropical residual soils. *12th South East Asian Geotechnical Conference, Kuala Lumpur, Malaysia* 1: 581-588

Powell, J.J.M. & Quarterman, R.S.T. 1988. The interpretation of cone penetration tests in clays, with particular reference to rate effects. *Proc. 1st Int. Symp. on Penetration Testing (ISOPT-1), Orlando, Fla.* 2: 903-909.

Premchitz, J. Gray, I. & Massey, J.B. 1988. Skin Friction on driven precast concrete piles founded in weathered granite. *Proc. 2nd Int. Conf. Geomechanics in Tropical Soils, Singapore* 1: 317-324.

Richardson, A.M., Brand, E.W. & Memon, A. 1975. In situ determination of anisotropy of a soft clay. *Proc. Spec. Conf. In Situ Measurement of Soil Properties, Raleigh, N.C., ASCE* 1: 336-349.

Serota, S. & Lowther, G. 1973. SPT practice meets critical review. *Ground Engineering* 6(1): 20-22.

Silvestri, V. & Aubertin, M. 1988. Anisotropy and in situ vane tests. In A.F. Richards et al. (eds), *Vane Shear Testing of Soils: Field and Laboratory Studies.* ASTM-STP 1014, Philadelphia, PA, pp. 88-103.

Silvestri, V., Aubertin, M. & Chapuis, R.P. 1993. A study of undrained shear strength using vari-ous vanes. *Geotechnical Testing Journal* 16(2): 228-237.

Sowers, G.F. 1985. Residual soils in the United States. In E.W. Brand & H.B. Phillipson (eds), *Sampling and Testing of Residual Soils*, a Review of International Practice, Technical Commit-tee on Sampling and Testing of Residual Soils. Scorpion Press, Hong Kong, pp. 183-191.

Stroud, M.A. 1974. The Standard Penetration Test in insensitive clays and soft rocks. *Proc. Euro-pean Symp. on Penetration Testing (ESOPT), Stockholm* 2(2): 367-375.

Thomas, D. 1965. Static penetration tests in London clay. *Géotechnique*, London, England XV(2): 177-187.

Vargas, M. 1974. Engineering properties of residual soils from south-central region of Brazil. *Proc. 2nd Int. Congr., IAEG, Sao Paulo* 1: 5.1-5.26.

Vaughan, P.R. 1988. Keynote paper: Characterising the mechanical properties of in situ residual soil. *Proc. 2nd Int. Conf. Geomechanics in Tropical Soils, Singapore* 2: 469-487.

Ward, W.H., Marsland, A. & Samuels, S.G. 1965. Properties of the London clay at the Ashford Common shaft: In situ and undrained strength tests. *Géotechnique*, London, England 15(4): 342-350.

Williams, A.A.B. 1980. Shear testing of some fissured clays. *7th Regional Conference for Africa on Soil Mechanics and Foundation Engineering, Accra, Ghana* 1: 133-139.

Wroth, C.P. 1984. The interpretation of in situ soil tests. *Géotechnique* 34(4): 449-489.

CHAPTER 10

Case histories of shear strength – controlled aspects of residual soils

G.E. BLIGHT
Civil Engineering Department, Witwatersrand University, Johannesburg, South Africa

10.1 THE STABILITY OF SLOPES IN RESIDUAL SOILS – FAILURE OF NATURAL SLOPES

Natural slopes in residual soil appear generally to be stable and to remain so unless conditions within the slopes become altered by unusual natural events or by human interference. Natural events that may precipitate slides in residual soil slopes are:

10.1.1 *Unusually heavy and prolonged rainfall*

The effects of unusually wet weather on the stability of slopes in residual soils have been examined by Lumb (1975), Morgenstern & de Matos (1975), Vargas & Pichler (1957), Da Costa Nunes (1969), Brand (1982), Malone & Shelton (1982), Pradel & Raad (1993), Lim et al. (1996) and Affendi & Faisal (1996). Many natural slopes of residual soil exist in an unsaturated condition and their margin of safety against sliding depends on the capillary tensions that exist in the pore water and enhance the strength of the soil. In a natural slope of weathered mudstone (Blight et al. 1970), the capillary tension was found to vary, after light rain, from zero at the surface to 1000 kPa at a depth of 1 m. Infiltration during prolonged rainfall can reduce capillary tensions to a point where the slope becomes unstable. Brand (1982) and Lim et al. (1996) have observed suctions in residual soil slopes in Hong Kong and Singapore. At shallow depths, suctions have been observed to decrease to zero during prolonged rain.

Lumb (1975), suggested that the limiting rate of infiltration of rain into a homogeneous soil in the absence of surface ponding is numerically equal to the saturated permeability of the soil. The water advances into the soil as a wetting front which travels at a velocity of $v = k/(1 - S)n$ where k is the permeability, S the initial degree of saturation and n the porosity. Using typical values for soil permeability and rainfall intensity, Lumb showed that such a wetting front could reach a potentially critical depth in a slope (such as the contact between soil and rock) within a few hours. Open cracks and fissures in a soil have the effect of accelerating the advance of such a wetting front.

Yong et al. (1982) have shown how varying duration of rainfall of an intensity at least equal to the saturated surface permeability reduces the factor of safety of a slope. Figure 10.1 shows some of the results of their analysis plotted as factor of safety versus depth of rainfall penetration. In the example, failure occurs after 20 hours in the one case and after 36 hours in the other. Rainfall of this sort of duration is by no means uncommon in many parts of the tropics, even though it may represent a 10 to 100 year event in other climatic areas.

One of the most extensive series of storm-associated slides on record was that which occurred in the Serra das Araras district of Brazil in 1967 (Da Costa Nunez

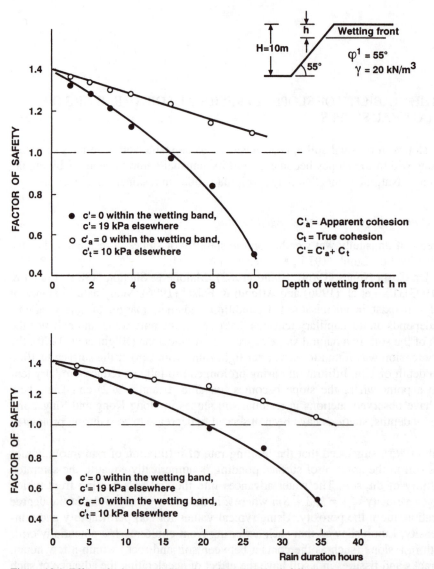

Figure 10.1. Effect of rainwater infiltration (above) and rainfall duration (below) on factor of safety of a slope.

1969). During and following a single night in which the rainfall intensity reached 70 mm/hour, an area 24 km long and 7.5 km wide was devastated by a series of landslides that killed an estimated 1000 people and caused untold damage to property.

Van Schalkwyk & Thomas (1991) record a similar set of circumstances in the KwaZulu Natal province of South Africa during 1987. Rainfalls totalling as much as 800 mm fell during a period of 4 days resulting in a loss of life of 380 and damage to property and infrastructure equivalent to 500 million US dollars. During this period, 211 slope failures occurred that damaged housing, roads and railways. In almost every case, the failures were associated with slopes in residual soils that had been subject to man's interference by cutting, terracing or constructing cut-to-fill road or railway alignments.

10.1.2 *Seismic events*

Yamanouchi & Murata (1973) describe a number of failures in slopes of shirasu (a residual volcanic soil) that occurred during the 1968 Ebino earthquake. As shirasu is a relatively rigid, brittle material, the earthquake loading produced multitudinous shear cracks in the soil which caused slabs of material to slough off. Although natural slopes appear to have been affected, cut slopes suffered more severe damage.

10.1.3 *Human interference*

There are many types of human interference that may affect the stability of natural slopes in residual soil. Of those the following are probably the most common:

1. Removal of toe support by cutting or erosion may precipitate a failure. The introduction of a cut at the toe of a long natural slope may precipitate a slide. The slide that occurred at Bethlehem, South Africa (Blight et al. 1970) is a typical example of this type of failure (see Fig. 10.2). The slide occurred when a shallow road cut was made in a hillside which slopes at a gentle 8°. On investigation it was found that the slide involved a block of weathered sand-stone that was sliding on its contact with an underlying stratum of weathered mudstone. Conditions for failure were exacerbated by a high water table and the existence at the sandstone-mudstone contact of a concentration of illite and montmorillonite clays leached out of the sandstone layer.

Material removed from the toe of a slope by erosion can also cause a previously stable slope to become unstable (e.g. Yamanouchi & Murata 1973).

2. Changes in the soil water regime of a slope may cause instability:

If the soil water regime of a slope is drastically altered by irrigation, the removal of vegetation or partial inundation by impounded water, instability may ensue (Richards 1985). The most spectacular and tragic recorded example of instability by raised water levels in a natural residual soil slope is the failure in the Vaiont valley described by Muller (1964) and Mencl (1966). During the impounding of water by a new dam, a slope some 600 m high, in residual weathered materials, forming one side of the reservoir valley began to creep. After creeping at varying rates for a period of over three years, during which certain observed points were displaced by 4 m, the sliding mass suddenly accelerated and the entire slope slumped into the reservoir, resulting in heavy loss of life and widespread devastation.

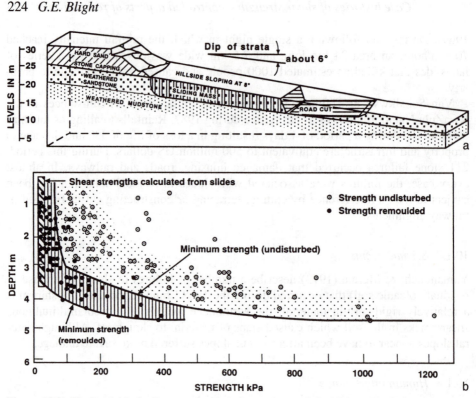

Figure 10.2) Sectional block diagram showing the geology of the slide at Bethlehem, b) Vane strength measurements in weathered mudstone at Amsterdamhoek.

3. The effect of deforestation may also affect the stability of natural slopes: The following appear to be the more important factors related to deforestation:

If deep-rooted vegetation is removed, capillary tensions will be reduced and the phreatic surface within a slope will probably rise (e.g. Blight 1987).

Roots mechanically reinforce the soil. The stability of a slope will decrease as the root systems decay after removal of surface vegetation. The relative importance of the above factors is dependent on the climatic environment. In a humid climate rotting of the root system may be rapid and decisive in producing instability. Under semi-arid conditions, however, the reduction in evapotranspiration rates caused by deforestation may outweigh all other effects.

10.2 FAILURES OF CUT SLOPES IN RESIDUAL SOILS

Many of the agencies that promote failure in natural slopes (unusually heavy rainfall, seismic activity, toe erosion, changes in soil water regime) may also effect the stability of cut slopes especially in the long term. However, the stability of cut slopes appears to be dominated by the structural features of the soil.

Residual soils that do not contain pronounced structural features such as relict jointing, bedding or slickensiding may be cut to remarkably steep stable slopes. For instance, Wesley (1973) cites several stable slopes cut by river erosion into a latosol clay. These slopes have heights of about 10 m and inclinations of up to 70°. He also mentions a 45° cut in similar material which is 20 m high and has been stable for over 40 years (see Fig. 10.3). Cut slopes in loess also show remarkable stability, stable vertical cuts of 5 to 6 m in height being quite common (Mitchell 1976). Woo et al. (1982) list come remarkable cut and natural slopes in cemented gravel deposits in north west Taiwan. Although these gravels are ancient alluvial deposits they have been laterized in situ and therefore qualify to be regarded as residual soils (Woo et al.'s observations are summarized in Fig. 10.4).

There is only one possible way in which the stability of the slopes illustrated in Figures 10.3 and 10.4 can be explained and that is by invoking the effects of pore water capillary tensions. Blight (1980) showed that a vertical slope with the water table at its toe can theoretically stand stably to any height if the capillary stresses within the slope are in static equilibrium with the water table. This is clearly how many such slopes remain stable over long periods of time.

Slides caused by the presence of joint surfaces, slickensides, etc. are very difficult to predict or design against as these features are generally not visible in borehole specimens and may be difficult to locate, measure or map even in test pits or trenches. Even if potentially dangerous features are identified during site or route exploration, it is usually practically impossible to assess their extent, frequency, variation in dip and other factors that would be needed to design a cut slope (e.g. Sandroni 1985).

St John et al. (1969), mention a number of cut failures attributable to the presence of structural weaknesses. In one series of incidents a 5½ km stretch of road traversing a mountainous volcanic tuff area of Puerto Rico suffered 40 separate slides in cuts ranging from 6 to 25 min depth. The slides were of two basic types which are

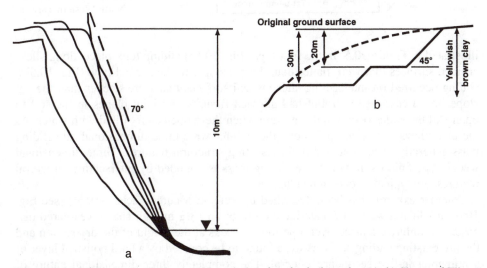

Figure 10.3. a) Cross sections along stable red clay river bank (latosol area), b) Slopes of Tjipanundjang dam borrow pit (andosol area) (after Wesley 1973).

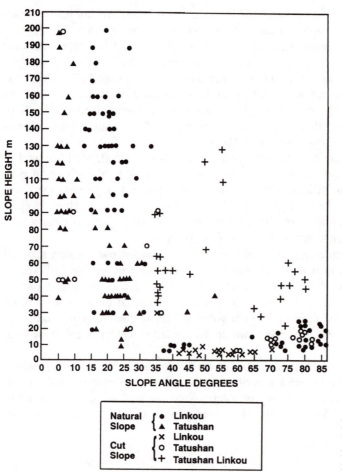

Figure 10.4. Slope height versus angle for slopes of ferricreted gravel deposits in North Western Taiwan.

Legend:
- Natural Slope — • Linkou, ▲ Tatushan
- Cut Slope — × Linkou, o Tatushan, + Tatushan Linkou

illustrated in Figure 10.5. In the first type (Fig. 10.5a) sliding took place along slickensided surfaces that were flatter than the cut slope. In the second type (Fig. 10.5b), sliding occurred on intersecting slickensided surfaces that were steeper than the cut slope. In all cases, stress relief had allowed fissures and joint planes in the soil to open. All the slides occurred during rain when these open clefts filled with water. As the plan views in Figure 10.5 show, the failures were three-dimensional, the sliding mass generally being bounded by intersecting structural features. Three-dimensional wedge-type failures, in which the sliding mass is bounded by intersecting structural features, are typical of cuts in residual soils.

Another example has been described by Pells & Maurenbrecher (1974) (see Fig. 10.6) in which a wedge of material slid out of an 18½ m high cut in weathered diabase. The failure occurred over a period of weeks at the height of the dry season and the pre-existing sliding surfaces were found to be covered by a hard polished layer of a horneblende-like secondary mineral. The completely three-dimensional nature of this failure is clearly shown by Figure 10.6.

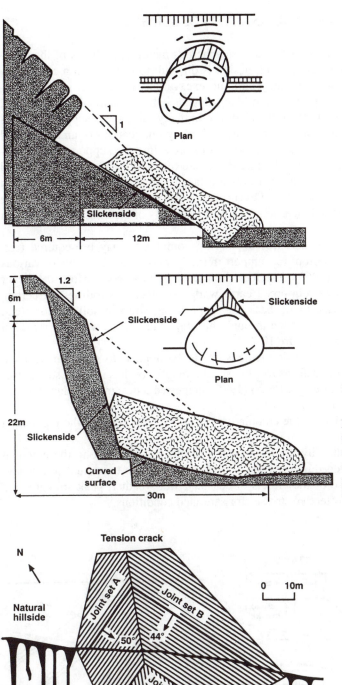

Figure 10.5. Slide on intersecting slickenside surface flatter (above) and steeper (below) than cut slope in weathered tuff (after St. John et al. 1969).

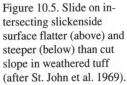

Figure 10.6. Plan showing three dimensional nature of slip in weathered diabase (after Pells & Maurenbrecher 1974).

10.3   DESIGN OF SLOPES AND ANALYSIS OF SLIDES

Because of the unknown and potentially treacherous nature and effects of the structural discontinuities in a residual soil, design is difficult to perform on a completely rational basis. For this reason, many building excavations in residual soil are completely supported, the support system being designed for a nominal coefficient of earth pressure. (e.g. Flintoff & Cowland 1982). In many cases these supported slopes would stand unsupported, but the economic and safety implications of an unexpected slide may be vastly more serious than the cost of providing the support. In less confined situations such as road or rail cuts, slopes are often cut to a nominal angle and if slides occur, ad hoc remedial measures are taken. As the surfaces exposed by a slide in residual soil are often stable even though steeper than the original cut face (e.g. St John et al.) remedial measures may consist simply of clearing away the fallen debris.

In other cases drainage measures of varying complexity may be required (e.g. Kezdi 1969) or the provision of toe support in the form of retaining walls of various types. Da Costa Nunes (1969) describes the use of gravity, counterfort, cantilever and crib walls. Gabion or reinforced earth walls are other possibilities. In certain cases it may be necessary to tie back the slope using stressed soil anchors (e.g. Wagener & Neely 1975). A section through this slide and one showing the remedial measures taken is shown in Figure 10.7.

If rational design and analysis methods are to be applied to slopes in residual soils, a decision has to be made as to what shear strength soil has in bulk, so that this value may be used in the design. Lumb (1975) suggested a design procedure for cuts in which it is assumed that either:
– Stability is controlled by the cumulative peak strength over the whole potential failure surface, or
– Stability is controlled by the least peak strength anywhere along the potential failure surface. This procedure provides bounds within which the actual soil behaviour should lie. Because of the effects of seasonal wet weather, peak strengths are established in terms of effective stresses for saturated conditions.

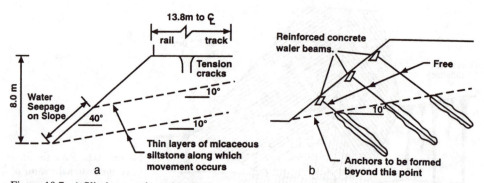

Figure 10.7. a) Slip in a cut in residual micaceous slitstone, b) Stabilization by means of stressed soil anchors (after Wagener & Neely 1975).

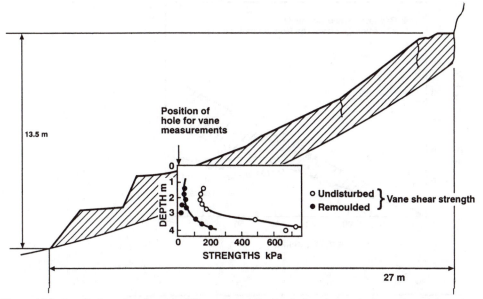

Figure 10.8. Detailed section through slide at Amsterdamhoek showing typical vane strength profile.

There is, however, considerable evidence (e.g. Burland et al. 1966, Blight 1969) that the shear strength of a stiff fissured material in bulk corresponds to the lower limit of strengths measured in small-scale laboratory or in situ tests (see, e.g. Figures 9.5, 9.41 and 9.42). This was shown by Blight (1969) by back analysing failures of four waste rock dumps founded on residual soil profiles. It was also demonstrated by Blight et al. (1970) in relation to series of slides that took place in a mantle of weathered cretaceous mudstone at Amsterdamhoek, South Africa (see Fig. 10.8). A section through the slide and a typical vane strength profile is shown in Figure 10.8. Figure 10.9 shows the shear strengths back-calculated from several slides at Amsterdamhoek plotted as a strength versus effective stress diagram. These values agree reasonably well with reversed shear box tests done in the laboratory. A similar principle can be shown to apply to three residual soils in Hong Kong by re-analysing data published by Malone & Shelton (1982). The data are shown in the form of strength diagrams in Figure 10.10. It will be noted that in each case the lower limit to the strength measured in the laboratory (triaxial shear) corresponds to the lower limit to strengths back figured from the slides. The analyses assumed zero pore pressure in the slopes ($r_u = 0$) whereas there was probably a small suction present. Hence the mean effective stress in Figure 10.10 has probably been under-estimated. If the lower limits to the strength measured in the laboratory had been used to analyse the various slopes a conservative but realistic result would have been obtained.

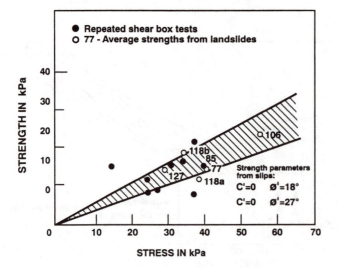

Location of Slide	Average shear strength kPa
House 77	15
House 85	16
House 106 (i)	24
House 106 (ii)	21
House 118 (i)	13
House 118 (ii)	19
House 127	15

Mean: 18 kPa

Figure 10.9. Residual strength characteristics of cretaceous rock (above) and average shear strengths calculated for five landslides at Amsterdamhoek (below).

10.4   TYPES OF FAILURE OF NATURAL SLOPES

Many attempts have been made to classify slides in natural slopes of residual soil (e.g. Vargas & Pichler 1957, and Morgenstern & de Matos 1975). Classification depends on two main characteristics – the geometry or shape of the sliding soil mass and the velocity of its motion.

10.4.1 *Geometry of slides*

There are two main types of slide geometry. If the residual soil mantle is shallow in comparison with the length of the slope, a planar slide may result. Commonly, the soil mantle slides over the underlying rock surface. The thickness of the sliding mass is usually roughly constant and the failure surface may be plane, slightly convex or slightly concave. The Caneleira slide described by Vargas and Pichler (see Fig.

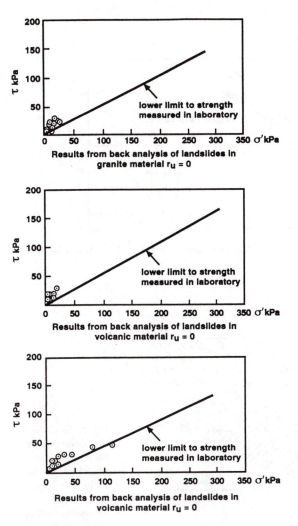

Figure 10.10. Comparison of lower limit to strength measured in laboratory with strengths back-figured from slope failures in three materials in Hong Kong (after Malone & Shelton 1982).

10.11) could be taken as the type planar slide. Other typical planar slides in natural residual soil slopes have been those at Tresna (Bujak 1967) in which the sliding surface was convex, Bethlehem (Fig. 10.2) and Amsterdamhoek (Fig. 10.8).

The second typical slide geometry occurs when the residual soil mantle is deep and a rotational or gouging slip or a block gliding occurs. The slide at Estrada do Jequia described by Morgenstern & de Matos (1975) typifies the rotational slide (Fig. 10.12), while the Danube slide described by Kezdi (1969) typifies the block gliding failure (Fig. 10.13).

10.4.2 *Slide velocity*

The velocity of a land-slide may vary from barely perceptible creep velocities to speeds of several kilometres per hour. Gray (1974) has recorded steady creep velocities in afforested slopes in Oregon, USA, averaging as little as 1 mm/year, while a

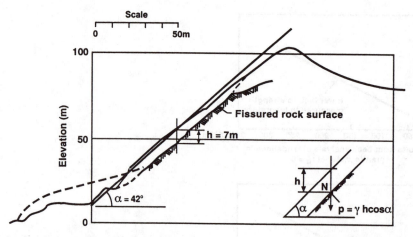

Figure 10.11. Section through a typical planar slide in residual soil at Canaleira, Brazil (after Vargas & Pichler 1957).

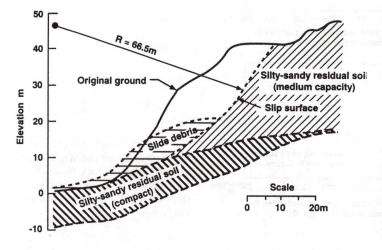

Figure 10.12. Section through a typical rotational slide in residual soil at Estrada do Jeqia, Brazil (after Morgenstern & de Matos 1975).

slope on the southern California coast quoted by Yen (1969) has been found to creep at 100 mm/year, and one in the Caucasus at 3 m/year (Ter-Stepanian 1965). Creep velocities are by no means constant with time, but vary seasonally and may be accelerated by wet weather and retarded by drought. An acceleration of the creep rate may lead to failure at a more rapid speed, as has happened in certain instances in Hong Kong (Lumb 1975) and at Amsterdamhoek (Blight et al. 1970) and Vaiont (Muller 1964).

At the opposite end of the velocity scale, flow slides may travel at much greater speeds. Data given by Lumb, for instance, suggests that velocities may approach 10 km/h, and it is possible that flow speeds may exceed this on steep slopes. Flow slides in residual soil that occur during heavy rain may travel considerable distances – a travel distance of over 5 km has been recorded by Da Costa Nunes (1969).

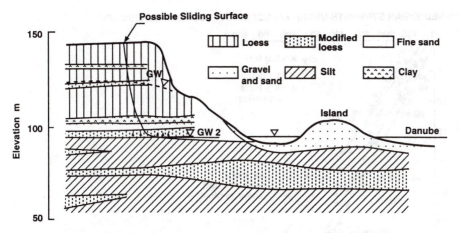

Figure 10.13. Section through unstable bank of river Danube (after Kezdi 1969).

10.5 SHALLOW FOUNDATIONS

The main problems with shallow, lightly loaded foundations on residual soils arise from seasonal or cumulative swelling or shrinkage or from collapse of highly leached, unstable grain structures or wetting. By their very nature, these problems tend to occur in the less humid tropical and sub-tropical areas where seasonal or perennial soil moisture deficits occur, usually in combination with active clays, highly weathered granites or loess-type soils. This topic has been dealt with in Section 8.4.

10.6 DEEP FOUNDATIONS

According to the Brand & Phillipson (1985) survey, deep foundations of various types are widely used in residual soils. Driven displacement piles and driven steel tube piles have been used in Brazil, but bored piles and caissons of various types appear to be more widely used in tropical soils. Hand dug caissons are widely used in Hong Kong, with bored contiguous piles frequently used to support the sides of building excavations. Bored piles are also used in India, Nigeria and Singapore. Driven H piles and precast concrete piles are used in Singapore. In South Africa the commonest type of pile in residual soils is the bored cast in situ pile, although driven and driven displacement piles are also used. The situation in Sri Lanka is similar.

When designing bored cast in situ piles in tropical soils, the question of what strength represents that of the soil in bulk must again be answered. Basically, the answer is the same as that given before – the lower limit to strengths measured by small scale means will represent the strength of the soil in bulk. However, in certain cases this is not a low enough value. For example, Figure 10.14 shows that strengths measured by means of large scale plug-pulling tests in an expansive residual siltstone (Blight 1984) were less than half of the lower limit to strengths measured by small scale means. Figure 10.15 shows a comparison of the predicted tension versus depth curve for a pile designed on the basis of the large scale shear tests, and tensions

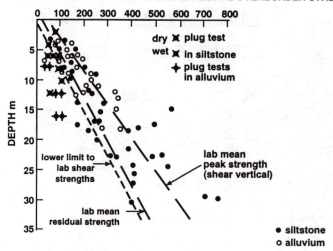

DRAINED SHEAR STRENGTH UNDER EFFECTIVE OVERBURDEN STRESS kPa

Figure 10.14. Shear strengths measured by means of large plugpulling tests compared with laboratory shear strengths on an alluvium and a residual siltstone.

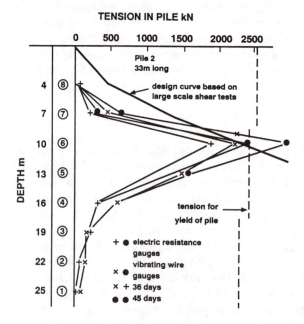

TENSION IN PILE kN

Figure 10.15. Comparison of predicted and measured tensions in a cast-in-situ pile installed in expansive residual siltstone.

measured in a test pile. The comparison shows that the large scale plug-pulling test results gave a completely realistic prediction of pile tensions, whereas the results of small scale tests would have seriously overestimated tensions.

Pavlakis (1983) has had considerable success in predicting pile failure loads and settlements from the results of pressuremeter tests in residual andesite lava. To predict the pile failure loads, he used standard Menard techniques, while to predict the load-settlement curve, he followed the procedure of Sellgren (1981). Figures 10.16a and 10.16b are examples of the excellent correlations he obtained between measured and predicted load-settlement curves (in Fig. 10.16a for a single driven cast-in-situ pile, and in Fig. 10.16b for two piles loaded through a single pile cap).

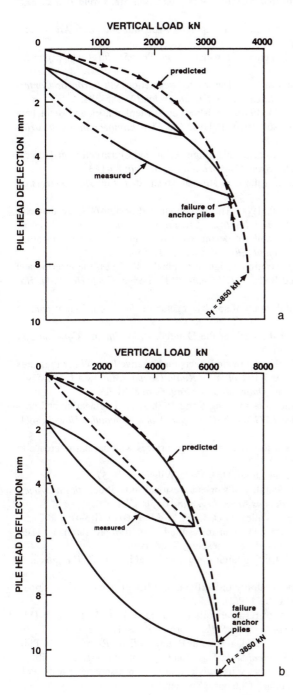

Figure 10.16. a) Measured and predicted load-deflection behaviour for a single cast-in-situ pile, b) Measured and predicted load-settlement behaviour of 2 driven cast-in-situ piles under one pile cap.

REFERENCES

Affendi, A. & Faisal, H.A. 1996. Suction, rainfall and slope stability. *12th South East Asian Geotechnical Conf. Kuala Lumpr, Malaysia* 1: 539-544.

Blight, G.E. 1969. Foundation failures in four rockfill slopes. *Jour. Soil Mech. and Found. Eng. Div., ASCE* 95(SM3): 743-767.

Blight, G.E. 1980. Partial saturation can assist the soil engineer. *Proc. 6th Southeast Asian Conf. on Soil Eng., Taipei, Taiwan* 1: 15-29.

Blight, G.E. 1984. Uplift forces measured in piles in expansive clay. *Proc 5th Int. Conf. Expansive Soils, Adelaide, Australia* 1: 240-244.

Blight, G.E. 1987. Lowering of the groundwater by deep rooted vegetation. *9th European Conference on Soil Mechanics and Foundation Engineering, Dublin, Ireland* 1: 285-288.

Blight, G.E., Brackley, I.J. & van Heerden, A. 1970. Landslides at Amsterdamhoek and Bethlehem – an examination of the mechanics of stiff fissued clays. *The Civil Engineer in South Africa* June, pp. 129-140.

Brand, E.W. 1982. Analysis and design in residual soils, Engineering and Construction in Tropical and Residual Soils, *ASCE, Geotech. Div. Spec. Conf., Honolulu, Hawaii*: 89-143.

Bujak, M. 1967. Preventive measures against the rock slide at Tresna dam site. *9th Congress on Large Dams, Istambul*: 1027-1033.

Burland, J.B., Butler, F.G.B. & Dunican, P. 1966. The behaviour and design of large diameter bored piles in stiff clay. *Symp. Large Bored Piles. Instn. Civ. Engrs., London*: 51-71.

Da Costa Nunes, A.J. 1969. Landslides in soils of decomposed rock due to intense rainstorms. *Proc. 7th Int. Conf. on Soil Mech. and Found. Eng., Mexico* 2: 547-554.

Flintoff, W.T. & Cowland, J.W. 1982. Excavation design in residual soil slopes. Engineering and Construction in Tropical and Residual Soils, *ASCE Geotech. Div. Spec. Conf. Honolulu, Hawaii*: 539-556.

Gray, D.H. 1974. Reinforcement and stabilisation of soil by vegetation. *Jour. of the Geotechnical Division, ASCE* 100(GT6): 695-699.

Kezdi, A. 1969. Landslide in loess along the bank of the Danube. *Proc. 7th Int. Conf. on Soil Mech. and Found. Eng., Mexico* 2: 617-626.

Lim, T.T., Rahardjo, H. & Change, M.F. 1996. Climatic Effects on Negative Pore-Water Pressures in a Residual Soil Slope. *4th Int. Conf. on Tropical Soils, Kuala Lumpur, Malaysia* 1: 568-574.

Lumb, P. 1975. Slope failures in Hong Kong. *Quarterly Jour. Eng. Geol* 8: 31-65.

Malone, A.W. & Shelton, J.C. 1962. Landslides in Hong Kong 1978-1980. Engineering and construction in tropical residual soils, *ASCE, Geotech. Div. Spec. Conf., Honolulu, Hawaii*: 424-442.

Mencl, V. 1966. Mechanics of landslides with non-circular slip surfaces with special reference to the Vaiont slide. *Géotechnique* 16(4): 329-337.

Mitchell, J.K. 1976. *Fundamentals of soil behaviour*. New York: Wiley, pp. 49-57.

Morgenstern, N.R. & de Matos, M. 1975. Stability of slopes in residual soils. Proc. *5th Pan. Amer. Conf. on Soil Mech. and Found. Eng., Buenos Aires, Argentina* 3: 369-384.

Muller, L. 1964. The rockslide in the Vaiont Valley. *Rock Mech. and Eng. Geol.* II(3-4): 148-212.

Pavlakis, M. 1983. Predictions of foundation behaviour in residual soils from pressuremeter tests. PhD Thesis, University of the Witwatersrand, Johannesburg, South Africa.

Pells, P.J.N. & Maurenbrecher, P.M. 1974. Cutting failures near Waterval Boven. *Civil Engineer in South Africa* 16(5): 180-181.

Pradel, D. & Raad, G. 1993. Effect of permeability on surficial stability of homogeneous slopes. *Journal of Geotechnical Engineering, ASCE* 119(2): 315-332.

Richards, B.G. 1985. Geotechnical aspects of residual soils in Australia. In E.W. Brand & H.B. Phillipson (eds), *Sampling and Testing of Residual Soils*. Scorpion Press, Hong Kong, pp 23-30.

Sandroni, S.S. 1985. Sampling and testing of residual soils in Brazil, In E.W. Brand & H.B. Phillipson (eds), *Sampling and Testing of Residual Soils*. Scorpion Press, Hong Kong, pp. 31-50

Sellgren, E. 1981. Friction piles in non-cohesive soils. Evaluation from pressuremeter tests. PhD Thesis, Chalmer's University of Technology, Goteburg, Sweden.

St John, B.J., Sowers, G.F. & Weaver, C.H.F. 1969. Slickensides in residual soils and their engineering significance. *Proc. 7th Int. Conf. on Soil Mech. and Found. Eng., Mexico City.*

Ter-Stepanian, G. 1963. *On the long term stability of slopes.* Norwegian Geotechnical Institute Publication 52: 1-13.

van Schalkwyk, A. & Thomas, M.A. 1991. Slope failures associated with the floods of September 1987 and February 1988 in Natal and KwaZulu, South Africa. In G.E. Blight (ed.), *Geotechnics in the African Environment.* Rotterdam: Balkema. Vol. 1, pp. 57-64.

Vargas, M. Pichler, E. 1975. Residual soil and rock slides in Santos, Brazil. *Proc. 4th Int. Conf. on Soil Mech. and Found. Eng., London* II: 394-398.

Wagener, F. von M. & Neely, W.J. 1975. Stability of a railway cutting in micaceous siltstones. *Proc. 6th Regional Conf. for Africa on Soil Mech. & Found. Eng.* 1: 213-218.

Wesley, L.D. 1973. Some basic engineering properties of halloysite and allophane clays in Java, Indonesia. *Géotechnique* 23(4): 471-494.

Woo, S.M., Guo, W.S., Yu, K. & Moh, Z.C. 1982. Engineering problems of gravel deposits in Taiwan. Engineering and Construction in Tropical and Residual Soils, *ASCE Geotech. Div. Spec. Conf. Honolulu, Hawaii:* 500-518.

Yamanouchi, T. Murata, H. 1973. Brittle failure of a volcanic ash soil – shirasu. *Proc. 8th Int. Conf. Soil Mech. and Found. Eng., Moscow* 1: 495-500.

Yen, B.C. 1969. Stability of slopes undergoing creep deformation. *Journal of the Soil Mechanics and Foundations Division, ASCE* 95(SM4): 1075-1096.

Yong, R.N., Siu, S.K.H. & Sciadas, N. 1982. Stability analysis of unsaturated soil slopes. Engineering and Construction in Tropical and Residual Soils, *ASCE Geotech. Div. Spec. Conf. Honolulu, Hawaii:* 483-499.